MW00610522

THE COMPLETE
STAR ATLAS

A PRACTICAL GUIDE TO VIEWING THE NIGHT SKY

Michael E. Bakich

Kalmbach
Media

Kalmbach Media
21027 Crossroads Circle
Waukesha, Wisconsin 53186
www.MyScienceShop.com

Published in 2020
24 23 22 21 20   1 2 3 4 5

Manufactured in China

ISBN: 978-1-62700-775-7
EISBN: 978-1-62700-776-4

The material in this book has appeared previously in *Astronomy* magazine. *Astronomy* is registered as a trademark.

Editor: Erica Barse
Book Design: Kelly Katlaps
Consulting Editor: Alison Klesman
Map Planning: Richard Talcott
Illustrator: Roen Kelly
Cover image: Tony Hallas

Library of Congress Control Number: 2019954800

The Complete Star Atlas
CONTENTS

Introduction

Welcome to what I hope will become your favorite pastime: stargazing. With minimal equipment and this thorough handbook, you can begin observing the extraordinary objects found in the heavens above.

To start you off, I've included descriptions of binoculars and telescopes for optimal night sky viewing. Our solar system alone offers many exciting points of interest, such as the Sun, the Moon, and the planets. Begin there, and then branch out into deep sky objects: double stars, galaxies, nebulae, and star clusters. I've also provided 50 observing tips that will hopefully guide you to a great viewing experience. The brightest stars are listed in the indexes, along with a guide to constellations and Messier and Caldwell objects (more on that later).

Finally, we get to the star charts. The cool thing about **The Complete Star Atlas** is that it's more than 24 terrific two-page star charts that cover the entire night sky. The descriptions, positional guides, and the many lists in the index come together as a beginning course in amateur astronomy. To those of you just starting your lifelong love affair with astronomy, view this atlas as a beginning, not an end. Let its information and images start you on a tour of the night sky's highlights. Along the way, you'll experience our fabulous universe.

Michael E. Bakich

 The icon at the upper left of each map shows the area of sky depicted; a label shows the map's central hour of right ascension.

Basics of
ASTRONOMY

BINOCULARS AND TELESCOPES • OBSERVING THE SOLAR SYSTEM •
OBSERVING THE DEEP SKY • OBSERVING TIPS • INDEXES

BINOCULARS AND TELESCOPES

Binoculars

Binoculars are versatile instruments with many benefits. They have a wide field of view and what you see through them is right-side up, making objects easy to find. They require no expertise to set up — just sling them around your neck and you're ready to go.

That portability also makes binoculars ideal for those clear nights when you might not have the time to set up a telescope. And for most people, observing with two eyes open rather than one seems more natural and comfortable. Most binoculars also are relatively inexpensive.

What the numbers mean

For stargazing, the size of the front lenses is the most important thing. The larger they are, the brighter the image will be. You can find the lens size simply by looking at the two numbers on every binocular: 7x35 or 10x50, for example. The second of those numbers refers to the size (in millimeters) of each front lens. So the front lenses of 7x35s have a diameter of 35mm, and 10x50s have a 50mm diameter. Binoculars with 50mm lenses gather twice as much light as 35mm binoculars.

Astronomy binoculars should have lenses at least 40 millimeters across. Smaller ones may be great for daytime viewing, but they won't gather enough light to give good views of most night sky objects.

The other number is the binoculars' magnification. For astronomy, go for binoculars that magnify at least seven times. The highest you'll want for handheld binoculars is about 10x. Too high a number here will over-magnify the motion of your hands, causing celestial objects to move around when you hold the binoculars. So, you likely won't be able to hold them steady enough to get a sharp image. High magnifications also limit the field of view, making objects more difficult to find for beginning observers. If you do choose a high-magnification model, use a binocular mount.

Binocular mounts

For the steadiest images possible, nothing beats mounting your binoculars to a tripod or custom binocular mount. Smaller, well-mounted binoculars with less magnification will, after a few short minutes of continuous use, beat hand-held binoculars of larger aperture and power.

The simplest binocular mount is an L-bracket. One end of it attaches to a mounting hole on the center post of the binoculars. The other end attaches to a camera tripod. Be sure your tripod is sturdy enough to carry the weight without shaking when you touch it.

Most amateur astronomers purchase commercially made mounts based on a design using a movable parallelogram. This keeps the binoculars pointed at an object over a wide range of motion, allowing people of varying heights to use them. It's ideal for observing sessions or star parties, where a number of people will be viewing the same objects.

If you do purchase a binocular mount, be sure it's rugged. One way to verify this is to center an object in the field of view and check whether the image settles down and shows no vibration after a few seconds (unless a strong wind is blowing).

▲ **THE AUTHOR BUILT** this parallelogram-style binocular mount. Because it has a movable counterweight, it can hold binoculars of all sizes and weights. MICHAEL E. BAKICH

Important details

Binoculars contain prisms that make the image appear right-side up. These prisms come in two varieties: roof and Porro. Roof-prism models have straight barrels and are more compact. However, they tend to be more expensive and produce slightly dimmer images, making them less desirable for astronomy. Porro-prism binoculars have a zigzag shape and usually are bigger and heavier than roof-prism models.

Lenses in high-quality binoculars are made of barium crown glass (BaK-4) instead of borosilicate glass (BK-7). Also, look for coated optics — the more lens and prism surfaces to which special coatings have been applied, the brighter and higher contrast the images will be.

Most binoculars have a central focusing knob that moves both eyepieces at once. These models also have one eyepiece that you can focus individually. To work the binoculars, first use the central knob to focus the eyepiece that doesn't adjust, and then focus the other eyepiece. This type is often more convenient to use, particularly if you pass the binoculars from person to person. On other binoculars, the eyepieces focus individually. These models tend to be more rugged and are often better sealed against moisture.

▲ **BINOCULARS USE** prisms to bend incoming light several times before it reaches your eyes. Manufacturers carefully position optical components so that no light is lost. HOLLEY Y. BAKICH

EXIT PUPILS

Exit pupil

An important term to know is *exit pupil*. This is the diameter of the shaft of light coming from each side of the binoculars to your eyes. If you point the front of the binoculars at a bright source, you'll see two small disks of light. These are images of the lenses. The diameter of the exit pupil (each disk of light) equals the aperture divided by the magnification. On a pair of 7x50 binoculars, the exit pupil diameter would be roughly 7 mm. For astronomical use, you want a large exit pupil because the pupils in our eyes dilate in darkness. The wider the shaft of light, the brighter the image will be because the light is hitting more of our retina. This loose rule, however, is only true up to a point. If the exit pupil is too large to fit into your eye, you lose some of the incoming light.

Some people have dark-adapted pupils measuring nearly 9 mm in diameter. Others have small ones less than 5 mm. Our pupils are largest when we're young. From age 30 on, they start to contract, slowing in our later decades. Additionally, women tend to have larger pupils than men of the same age, on average. Unfortunately, there's no hard-and-fast rule that correlates pupil size with age or gender. You can measure your pupil size with a gauge available from some astronomy equipment suppliers or from an ophthalmic or pharmaceutical company.

EXIT PUPIL
This is the diameter of the shaft of light coming from each side of the binoculars to your eyes.

Eye relief

Eye relief is the manufacturer's recommended distance your pupil should be from the eyepiece lens for best performance. Eye relief generally decreases as power increases. Eye relief less than 10 mm requires you to get uncomfortably close to the eyepieces. This is no problem for advanced amateurs, but for beginners, higher eye relief allows the head more freedom of movement. Also, those who must wear eyeglasses (for example, to correct for astigmatism) need more eye relief.

◀ **THE ILLUMINATED openings of these binoculars are the exit pupils.** MICHAEL E. BAKICH

▲ IF AT ALL POSSIBLE, handle the binoculars before you buy them. This will help you assess their workmanship and also let you know how much they weigh. COURTESY OF CELESTRON

/// **BUYING BINOCULARS**

Use these five helpful guidelines to select binoculars:

1 Pick up the binoculars and shake them gently. Then twist them gently. Move the focusing mechanisms several times and move the barrels together, then apart. Look for quality of workmanship. If you hear loose parts or if there's any play, don't buy them. Another consideration is the weight of the binoculars. If you're going to be hand-holding them, try to imagine what they will feel like at the end of an observing session.

2 Look into the front of the binoculars and check for dirt or other contaminants. Apart from perhaps some dust on the outside of the lenses, the inside should be immaculate.

3 Hold the binoculars in front of you with the eyepieces toward you. Point them at a bright area. You will see the disks of light that are the exit pupils. They should be round. If not, the alignment of the binoculars is bad and the prisms are not imaging all of the light.

4 Test the binoculars on a target. Try to do this outdoors and at night. Nothing reveals flaws more than star images. If you can't test the unit at night or even outdoors, try to look through a door or window at distant objects. How well do the binoculars focus? Are objects clear? If there is any sign of a "double" image, the two barrels are not aligned. If you are wearing glasses, can you get your eyes close enough to the binoculars to see the entire field of view?

5 Finally, repeat tests 1–4 with several different binoculars. Once you become more familiar with how they compare, you'll be well on your way to purchasing an excellent unit.

A note about caring for your binoculars: Generally, binoculars come with lens caps, eyepiece caps, and a case. Use them. These protect your binoculars from dust and moisture. Don't leave your binoculars exposed to direct sunlight, even in their case. And keep the vibrations (especially impacts) to a minimum.

◄ **CELESTRON'S SKYMASTER**
25x100 binoculars fall into the "giant" category. Each barrel is essentially a 4-inch telescope.

COURTESY OF CELESTRON

What you'll see

Binoculars will show the Moon in crisp detail. Watch shadows creep across lunar features as the Moon's phase changes. Follow the stages of a lunar eclipse as Earth's shadow covers the Moon. And view a crescent Moon silhouetted against stars low in the western evening sky.

Farther afield, binoculars let you track Jupiter's four big moons. They'll help you pick out Mercury low in the twilight sky and spot objects too faint to see easily, such as the ice giant planets, Uranus and Neptune, as well as the brighter asteroids.

The advantages of binoculars perhaps show up best when viewing a bright comet. Binoculars magnify enough to show exquisite detail while providing a wide enough field of view that you can see the comet's head and most or all of its tail at once.

Binoculars also put many of the deep-sky targets in this atlas within your grasp. Star clusters, especially large ones like the Pleiades (M45), the Salt-and-Pepper Cluster (M37), and h and Chi (χ) Persei (NGC 869 and NGC 884) look particularly good through them. And what better way is there to relax at a star party or observing session than by reclining in a chair and scanning the Milky Way through binoculars?

▼ **IMAGE-STABILIZED** binoculars, like this 18x50 model made by Canon, reduce the effect of shaky hands.

MICHAEL E. BAKICH

Buying a telescope is a big step, especially if you're not sure what all those terms like f/ratio, go-to, and magnification mean. To help you understand what to look for, let's answer 10 of the most-asked questions.

1 What exactly does a telescope do?

A telescope's purpose is to collect light. This property lets you observe objects much fainter than you can see with your eyes alone. Galileo said it best when he declared that his telescopes "revealed the invisible."

2 Will my telescope be complete, or will I need more to make it work?

Many telescopes are complete systems, ready for the sky as soon as you unpack and assemble them. But a few models are "optical tube assemblies," or OTAs. This means all you're buying is the optics and the tube with no tripod or accessories.

3 I want to observe. What should I do first?

Your goal should be to learn all you can about telescopes: what types are available, what accessories are the best, and what you'll see through them.

If a telescope interests you, research different types on the internet. Look specifically for telescope reviews. You'll learn what's important to veteran observers when they use a telescope. You'll also get a feel for optical and mechanical quality, ease of use (including portability), and extra features.

4 Why are objects in my telescope upside down?

As light enters the eyepiece, the top of what you're looking at is at the bottom, and vice versa. You can re-flip the image with an accessory called an "image erector," but you'll lose some of the object's light. And in observing, you want your telescope to deliver the maximum amount of light possible to your eye. (Besides, there's no up or down in space.)

5 Can I use my telescope for views of earthly objects?

Absolutely! Many nighttime observers (usually those with smaller telescopes) also use their telescopes for bird-watching or other nature activities. Here's where an image erector comes in most handy.

6 Is there a way for me to "test-drive" a telescope?

Yes. Many areas have astronomy clubs and you can attend a meeting. There, you'll find others who enjoy the hobby and are willing to share information and views through their telescopes. At one of the club's stargazing sessions, you'll be able to look through many different telescopes in a short period and ask all the questions you like.

7 Is a "go-to" scope better than one without go-to?

Yes. A go-to scope is one with a motor or motors controlled by a computer. Once set up, a go-to scope will save you lots of time by moving to any sky object you select and then tracking it. Even experienced observers prefer go-to scopes because they leave more time to observe the sky.

8 Apart from the optics, what's the most important thing in a telescope system?

The most crucial component is the mount, which is what the telescope's tube sits on. You can buy the finest optics on the planet, but if you put them on a low-quality mount, you won't be happy. No telescope can function in high winds, but a poor mount will transfer vibrations even in a light breeze.

© VNLIT | DREAMSTIME.COM

9 Does a telescope need electricity?

Only if it has a motorized drive. In most cases, telescope drives use direct current, which means you can use batteries (including the one in your car). Adapters will let you plug your scope into an electrical outlet.

10 What is the best telescope for me?

It's the one you'll use the most. If it takes an hour to set up a scope, or if your scope is difficult to move, you might observe only a few times each year. If your scope is quick to set up, you may use it several times each week. A small telescope that's used a lot beats a big scope collecting dust in a closet every time.

Refracting telescopes

The word *refract* means "to bend." A refracting telescope (or refractor) does this with a carefully made lens system. If the surfaces of the lenses have the proper shape, the light will come to a focus. Placing an eyepiece at the focus will let you see the target.

Dutch eyeglass maker Hans Lippershey made the first telescope (a refractor) in 1608. Italian inventor Galileo Galilei was the first to use the telescope to study celestial objects, and what he saw revolutionized astronomy forever.

Two words you'll see when reading about refractors are *achromat* and *apochromat*. The earliest telescopes had poor lenses with various defects. In 1729, the first lens that combined two different types of glass appeared. This type of lens is an "achromat," which means "not color dependent." An achromatic lens does a good job of bringing all colors of light to the same focus.

By the 1980s, even better "apochromatic" lenses called became widely available for consumer telescopes. Apochromats may use two lenses like an achromat, but they're more likely to have three or four. The main difference between the two types is the amount of color fringing you'll see on bright objects. It usually appears as a purple border on one side of the object. Less fringing is better.

One selling point of a refractor is that nothing blocks or scatters any of the light passing through the lens, which makes image contrast better. Observers of planets and double stars say that refractors are best for such objects.

Refractors also are low maintenance. Lenses never require recoating like mirrors do. Also, a lens usually doesn't need adjustment. It won't get out of alignment unless the scope encounters a major trauma.

One slight downside is a refractor's closed tube, which requires time to adjust to the outside temperature when moved from a warm or cool house to the outdoors. Today's thin-walled aluminum tubes have reduced cool-down time a lot, but you still have to take it into account.

▼ **A REFRACTOR USES a lens (a combination of two to four glass pieces) to bring light to a focus.** *ASTRONOMY:* ROEN KELLY, AFTER CELESTRON

Lens shade

Eyepiece

Telescope tube

Light enters here

Lens

Focuser

▶ **CELESTRON'S NEXSTAR 102SLT** package contains a 4-inch refractor on a computerized mount.
COURTESY OF CELESTRON

◀ **CELESTRON'S POWERSEEKER 60AZ** is an example of a small, low-priced refractor. This model has a 2.4-inch lens, sits on a stable mount, and produces right-side-up images with the supplied diagonal. COURTESY OF CELESTRON

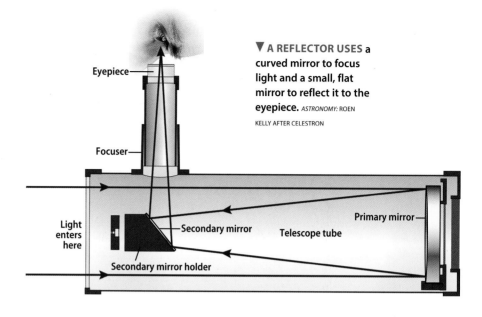

▼ A REFLECTOR USES a curved mirror to focus light and a small, flat mirror to reflect it to the eyepiece. *ASTRONOMY: ROEN KELLY AFTER CELESTRON*

Eyepiece

Focuser

Light enters here

Secondary mirror

Secondary mirror holder

Telescope tube

Primary mirror

◄ CELESTRON'S ASTROMASTER 130EQ is a 5.1-inch reflector on an equatorial mount. COURTESY OF CELESTRON

Reflecting telescopes

Scottish astronomer James Gregory invented the reflecting telescope and published a description of it in 1663. Although he's given credit for the invention, Gregory never actually made one.

English mathematician Sir Isaac Newton constructed the first working reflector in 1668. It had a mirror 1 inch across and a tube 6 inches long. A Newtonian reflector (still the main type sold today) contains two mirrors: one called the "primary" at the bottom of the tube, and a small, flat "secondary" near the top of the tube. Light enters the top, travels down the tube, hits the primary, and reflects to the secondary. That mirror then reflects it into the eyepiece.

One benefit is that a reflector doesn't create color fringes around even the brightest objects. Another selling point is that reflectors are less expensive than refractors. When working with a mirror, manufacturers have to polish only one surface. An apochromatic lens has between four and eight surfaces — plus, you're looking

through the lenses, so the glass has to be defect-free. All of this makes such lenses a lot more expensive.

There are a few negatives to reflectors. The placement of the secondary mirror creates an obstruction that scatters a tiny amount of light from bright areas into darker ones. For most objects, you'll never notice this.

Newtonian reflectors suffer from "coma," a defect where stars at the edge of the field of view look long and thin like a comet. Observers compensate for this by placing targets at the center of the field.

Finally, because of how the mirror attaches to the tube, a reflector is sensitive to vibrations when transported. Observers usually collimate, or adjust the mirrors in their telescope, before each observing session.

▲ CELESTRON'S SKYPRODIGY 130 is a reflector that contains a 5.1-inch primary mirror. COURTESY OF CELESTRON

Compound telescopes

The word *catadioptric* means "due to both the reflection and refraction of light." These instruments also are known as compound telescopes and are hybrids that have a mix of refractor and reflector elements in their design.

German astronomer Bernhard Schmidt made the first compound telescope in 1930, which was the precursor of today's most popular design, the Schmidt-Cassegrain telescope, or SCT. In this instrument, light enters the tube through a glass corrector plate and then hits the primary mirror at the tube's base, which reflects the light to a secondary mirror mounted on the corrector. The secondary reflects light through a hole in the primary mirror to the eyepiece, which sits at the back of the scope.

The number-one advantage of an SCT is its compact design. Such instruments are often only one-quarter as long as comparably sized reflectors or refractors. This makes the SCT one of the ultimate grab-and-go telescopes.

The negative is that, like refractors, compound telescopes have a closed tube. Adjusting to the outside temperature, therefore, takes longer than with an open-tube reflector of the same size.

▼ **A COMPOUND TELESCOPE** combines a front lens with mirrors to focus light. This diagram shows a Schmidt-Cassegrain telescope.
ASTRONOMY: ROEN KELLY, AFTER CELESTRON

▲ **CELESTRON'S NEXSTAR 127SLT is a 5-inch compound telescope supplied with a go-to mount that runs on eight AA batteries (or an optional adapter).** COURTESY OF CELESTRON

Primary mirror

Telescope tube

Light enters here

Eyepiece

Focus knob

Secondary mirror

Corrector plate

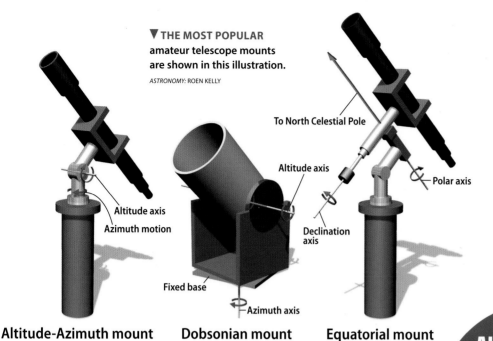

▼ **THE MOST POPULAR** amateur telescope mounts are shown in this illustration.
ASTRONOMY: ROEN KELLY

Altitude axis
Azimuth motion

To North Celestial Pole

Altitude axis

Polar axis

Declination axis

Fixed base

Azimuth axis

Altitude-Azimuth mount   **Dobsonian mount**   **Equatorial mount**

Mounts and drives

We call these instruments telescopes, but we could say, "optical tube on a mount." Half of any telescope is its mount, and some observers claim that the mount is the more important part. If the mount is too light, wind will be only one of your enemies. Your images will bounce around even when you are focusing.

An alt-azimuth mount is the simplest type of telescope mount. The name is a combination of "altitude" and "azimuth." A telescope on this type of mount moves up and down (altitude), and left and right (azimuth).

In the 1960s, amateur astronomer John Dobson invented a simple type of alt-azimuth mount that now bears his name. The Dobsonian mount is the least expensive telescope mount and manufacturers always combine it with a reflecting telescope. Because the telescope tube sits loosely in the mount, an observer can carry the two parts quite easily.

ALT-AZIMUTH MOUNT
This is the simplest type of telescope mount. The name is a combination of "altitude" and "azimuth."

► **CELESTRON'S ASTROMASTER** tripod is a simple alt-azimuth assembly on which you can mount either binoculars or a small telescope.
COURTESY OF CELESTRON

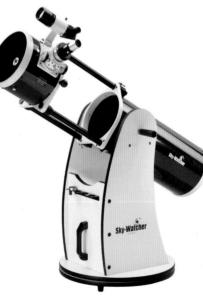

▲ **CELESTRON'S SKY-WATCHER** Dobsonian line combines a Newtonian reflector with an easy-to-use Dobsonian mount.
COURTESY OF CELESTRON

▲ **CELESTRON'S NEXSTAR 6SE** package combines a 6-inch Schmidt-Cassegrain telescope, a tripod, and a computerized go-to mount. COURTESY OF CELESTRON

One development in the past few decades is the go-to mount. To create this, manufacturers attach motors connected to a computer to a telescope's altitude and azimuth axes. Once you run through a simple setup procedure, the go-to drive will find and then track your celestial target. Most go-to scopes manufactured today have large, built-in databases featuring thousands of objects.

▲ **CELESTRON'S FINDER SCOPE** kit features a finder scope with a 2-inch (50 mm) front lens and a magnification of 9×. COURTESY OF CELESTRON

▲ **THIS SMALL FINDER SCOPE** doesn't magnify. Rather, it projects a red dot onto a transparent screen.
COURTESY OF CELESTRON

▲ **A STAR DIAGONAL** bends light 90°. This accessory makes observing a lot more comfortable. COURTESY OF CELESTRON

SKY POSITIONS AND COORDINATES

Imagine the sky as a sphere of infinite size, centered on Earth. This concept works because distances to celestial objects are not discernible to the eye, so objects appear to lie on a great sphere far away. Astronomers use two coordinate systems to locate these objects.

The alt-azimuth coordinate system

In this system, **altitude** is the number of degrees from the horizon to the object. It ranges from 0° (horizon), to 90° (zenith).

We measure **azimuth** along the horizon from north through east to where a line passing through the object intersects the horizon at a right angle. Azimuth varies between 0° and 360°.

A brief note about angles: A circle contains 360°. Each degree is divided into 60 arcminutes, and each arcminute is divided into 60 arcseconds. The symbol ' designates an arcminute and " indicates an arcsecond.

The alt-azimuth system's main disadvantage is that objects' coordinates change constantly because of Earth's rotation. We solve this problem by fixing coordinates to the celestial sphere.

The equatorial coordinate system

Imagine projecting Earth's equator and poles to the celestial sphere. This produces the celestial equator as well as the North and South Celestial Poles.

Declination corresponds to Earth's latitude and is the angle between the object and the celestial equator. It varies from 0° to 90° north or south and is measured in degrees (°), arcminutes ('), and arcseconds ("). A minus sign (–) is used for objects south of the celestial equator.

Circles that run through the celestial poles perpendicular to the celestial equator are hour circles. To designate the position of a star, consider one of these great circles passing through the celestial poles and through the star. This is the star's hour circle, and it corresponds to a meridian of longitude on Earth.

All that's left is to set the zero point of the "longitude" coordinate, which is called **right ascension**. For this, astronomers use the vernal equinox, an intersection point of Earth's equator and its orbital plane, the ecliptic. The Sun appears to move through this point each year around March 21, moving from south to north crossing the celestial equator.

The angle between the vernal equinox and the point where the object's hour circle intersects the celestial equator is the object's right ascension. It is measured in hours (h), minutes (m), and seconds (s). Right ascension is measured from west to east and begins at 0h (the vernal equinox). Each hour corresponds to 15°.

Earth's motions

Earth's rotation has remarkable effects on the sky's appearance: Celestial objects appear to circle the celestial poles. If a star's distance from the pole is greater than your latitude, you won't see the entire circle. Stars simply will rise in the east, move across the sky, and set in the west. If that star is circumpolar for your latitude, you'll see its entire circle.

Earth's revolution around the Sun causes seasonal changes. The effect on the night sky is a slow westward progression of the constellations throughout the year.

For example, if you go out tonight and look at the stars' positions at, let's say, 8 P.M., then tomorrow night, the stars will be in the same position 4 minutes earlier, at 7:56 P.M. The following night, they'll reach that position at 7:52 P.M., and so on. A bit of math reveals that in a month, the stars will be 2 hours out of sync with your first observation.

So, if on your first night you saw Gemini the Twins rising low in the east at 8 P.M., a month later they'll be much higher at that time. And in three months — one season — Gemini (and all the other stars) will have moved a quarter of the way across the sky. Four seasons (one year) later, they'll be back in the position where you first saw them.

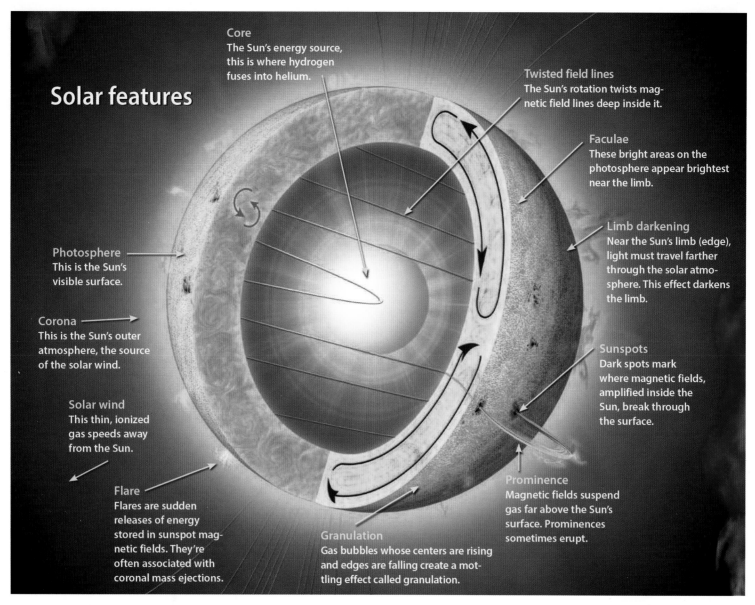

Solar features

Core
The Sun's energy source, this is where hydrogen fuses into helium.

Twisted field lines
The Sun's rotation twists magnetic field lines deep inside it.

Faculae
These bright areas on the photosphere appear brightest near the limb.

Limb darkening
Near the Sun's limb (edge), light must travel farther through the solar atmosphere. This effect darkens the limb.

Photosphere
This is the Sun's visible surface.

Corona
This is the Sun's outer atmosphere, the source of the solar wind.

Solar wind
This thin, ionized gas speeds away from the Sun.

Sunspots
Dark spots mark where magnetic fields, amplified inside the Sun, break through the surface.

Flare
Flares are sudden releases of energy stored in sunspot magnetic fields. They're often associated with coronal mass ejections.

Granulation
Gas bubbles whose centers are rising and edges are falling create a mottling effect called granulation.

Prominence
Magnetic fields suspend gas far above the Sun's surface. Prominences sometimes erupt.

▲ **THE SUN** is the closest star to Earth. Several observable features arise from its complex physics.
ASTRONOMY: ROEN KELLY

OBSERVING THE SOLAR SYSTEM

Observing the Sun

The Sun, because it's the sky's brightest object, also is the easiest to observe. Put safety first and even a small telescope will delight you with high-quality views. Plus, you can make good solar observations even when conditions would rule out seeing other celestial objects.

Start with the disk

The **PHOTOSPHERE** is the Sun's visible surface and is the lowest observable layer of the solar atmosphere. Observing the photosphere is easy through visible-light solar filters.

If the visibility is particularly good, you'll spot **GRANULATION**, which observers describe as giving the photosphere a mottled effect, with

both light and dark areas. Vast gas bubbles create the granules as their centers rise and their edges sink.

FACULAE are bright areas visible on the photosphere. *Facula* is Latin for "little torch." Faculae appear all over the solar disk, but observers most often see them near the solar limb (edge). There, the contrast between the

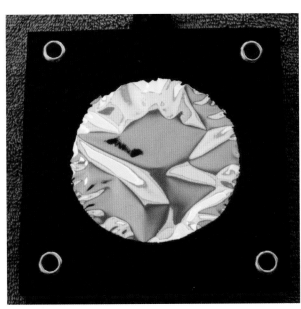

▲ **AN APPROVED VISIBLE-LIGHT** solar filter will let you see sunspots and several other features. MICHAEL E. BAKICH

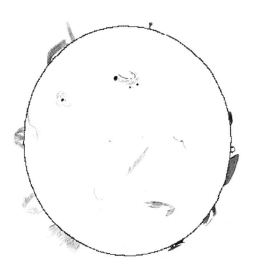

▲ **SKETCHING THE SUN** through a Hydrogen-alpha filter reveals prominences and solar flares. MICHAEL E. BAKICH

faculae and the darkened limb is highest.

Finally, look for a phenomenon called **LIMB DARKENING**. We observe limb darkening because the Sun is a sphere. Near what we see as the edge of the solar disk, the light must travel farther through the solar atmosphere. This causes the limb to be dimmer than the rest of the disk.

The chromosphere

The chromosphere, or "sphere of color," lies just above the photosphere. Here, hydrogen atoms emit energy called Hydrogen-alpha (Hα) radiation. Hα is reddish-colored light with a wavelength of 656.28 nanometers (nm).

Through an Hα filter, which allows only Hα light through, you'll see **PROMINENCES**, bright gas clouds ejected from the Sun and shaped by its magnetic field. Prominences appear as spikes, loops, detached regions, and more. Sometimes prominences look like dark lines when they are silhouetted against the brighter solar disk; astronomers call these

filaments. Plages are another Hα feature that appears as bright areas around sunspots.

Solar explosions

Also best seen through Hα filters, solar **FLARES** occur when the Sun's atmosphere suddenly releases built-up magnetic energy. Solar flares emit radiation storms and are the solar system's largest explosions. They appear as bright regions on the disk.

Flares can be classified by how much area they cover at maximum brightness. They range from Importance 0, or subflares (smaller than 2 square degrees), to Importance 4, which cover more than 24.8 square degrees. On the Sun, one square degree equals roughly 57 million square miles (150 million square kilometers).

Dark areas in a sea of light

SUNSPOTS, which are features of the photosphere, come in many shapes and sizes, based on the activity of the Sun's magnetic field. The field traps gas, slowing its motion and making it cooler and darker than the surrounding area.

Usually, sunspots consist of a dark central region called the umbra, surrounded by a lighter region known as the penumbra. The penumbra's temperature is typically 1,800° Fahrenheit (1,000° Celsius) below that of the photosphere, and the umbra is usually between 2,700° F (1,500° C) and 3,600° F (2,000° C) cooler than the photosphere.

Roughly every 11 years, solar activity peaks, resulting in greater numbers of sunspots and flares. German astronomer Samuel Heinrich Schwabe discovered the sunspot cycle in 1843. This cycle varies from as few as 9.5 to as many as 12.5 years. The start of any given solar cycle is defined as the minimum of sunspot activity. Since the 19th century, astronomers have recorded sunspot numbers each day.

► THIS SIMPLE SETUP **uses binoculars to project two images of the Sun.**
MICHAEL E. BAKICH

Observing by solar projection

One way to observe the Sun safely is by using an attachment to project the Sun's image. Some observers use an adjustable arm that holds a sheet of paper behind their eyepiece. Others use a box assembly. A box is a better choice because it darkens the surrounding area and increases contrast. Whichever you use, before you start, draw on a sheet of paper a 6-inch-wide (150 mm) circle — the standard size used by observers worldwide.

Mark the four directions, focus the Sun, and fit it to your circle. If it doesn't fit, either adjust the eyepiece/paper distance or choose an eyepiece with a different focal length. Don't use older eyepieces whose lens elements are held together by cement, because the Sun's heat will damage them.

► DURING A TOTAL **solar eclipse, the corona, the Sun's delicate outer atmosphere, can be seen.** NASA/AUBREY GEMIGNANI

Visual solar filters

A good solar filter is safe — it does not transmit harmful ultraviolet or infrared radiation, and it drops the Sun's brightness to a comfortable level.

Visible-light filters are either coated glass or optical-quality Mylar. The solar image through Mylar usually appears white or bluish-white; through glass filters it can appear white, yellow, or orange. Glass filters are more expensive but also more durable.

All solar filters fit over a telescope's objective (front) end. Some cover the entire objective (full-aperture filters), while others have smaller openings offset from center (off-axis filters). Off-axis filters eliminate secondary-mirror obstructions. All solar filters should have round openings. Other shapes introduce distracting diffraction patterns.

Never use a solar filter that fits into an eyepiece. Some of these filters can crack due to heat buildup, with devastating results.

NEVER use a solar filter that fits into an eyepiece. Some of these filters can crack due to heat buildup, with devastating results.

Hydrogen-alpha (Hα) filters

Observing the Sun at the wavelength of Hα light is one of amateur astronomy's fastest-growing segments. All Hα filters center on a wavelength of 656.28 nm. However, such filters have different bandpass widths. The widest of these can be nearly 2 Angstroms (Å) and the narrowest 0.3 Å.

One Angstrom equals 0.1 nanometer. Through a 1 Å-band-pass Hα filter, prominences look great but chromospheric detail is low. Through a filter with a bandpass of 0.5 Å, you'll see lots of chromospheric detail but few prominences. Some Hα filters are tunable; you can shift the bandpass' central wavelength slightly to either side.

Solar observing is addictive. Soon, you'll find yourself watching the Sun as much as you do the stars. Don't forget the sunscreen.

Observing the Moon

The Moon offers something for every amateur astronomer. It's visible somewhere in the sky most nights, its changing face presents different features from one night to the next, and it doesn't take an expensive setup to enjoy it. To help you get the most out of viewing the Moon, I've compiled these simple tips. Follow them, and you'll be well on your way to a lifetime of satisfying lunar observing.

Perform a "no optics" survey

The best way to begin your journey as a lunar observer is to learn the Moon's major features. Head out with a simple Moon map and use just your eyes to identify our only natural satellite's top attributes.

The Full Moon is for romance

Contrary to what you might think, Full Moon is not the best time to observe our natural satellite, although that's the time when it's at its brightest. When the Moon is full, the Sun lies behind Earth (as we face the Moon), shining directly onto the lunar surface. Shadows are at their minimum lengths and you can't see much detail. You can still observe the Moon when it's full, but the contrast between its light and dark sections will be much better at other times.

▲ **THE FULL MOON** may be pretty to look at, but it's the worst time to actually observe our lone natural satellite. NASA

▶ **OBSERVE WHEN** the Moon is not full (left). At Full Moon (right), the Sun is directly overhead and shadow detail is reduced. HOLLEY Y. BAKICH

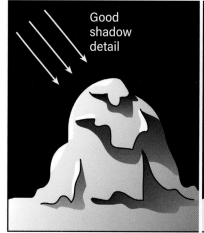

Good shadow detail

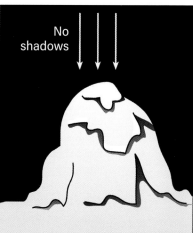

No shadows

View the Moon at "prime time"

Two intervals during the lunar "month" (from one New Moon to the next) are best for observers. The first begins shortly after New Moon and continues until two days past First Quarter. Amateur astronomers tend to favor this span because the Moon lies in the evening sky. An equally good observing period starts about two days before Last Quarter and goes until the Moon lies so close to the Sun that it's lost in morning twilight. At these times, shadows are longer and features stand out in sharp relief. Another benefit you'll get when you observe the Last Quarter Moon is that the atmosphere before dawn is usually steadier than it is after sunset. When the Sun sets, a lot of heat remains in the atmosphere. As hot air rises and cooler air sinks, the resulting turbulence leads to unstable air — what observers call bad seeing.

The terminator will help you

During the two favorable periods described above, point your telescope anywhere along the line that divides the Moon's light and dark portions. Astronomers refer to this line as the "terminator." Before Full Moon, the terminator marks where sunrise is occurring. After Full Moon, sunset happens along the terminator. Here you'll spot the tops of mountains protruding just high enough to catch the Sun's light while surrounded by lower terrain that remains in shadow. Features along the terminator change in real time and, during a night's observing, the differences you'll see through your telescope are striking.

The best scope for viewing the Moon

Nearly any telescope will do to observe lunar details. You'll get great views of the Moon through a 2.4-inch refractor, an 8-inch reflector, or an 11-inch Schmidt-Cassegrain telescope. Observers with several options (but not a permanent observatory) usually pick a scope they can set up many nights in a row. Observing on successive nights makes it easier to follow the terminator's progress.

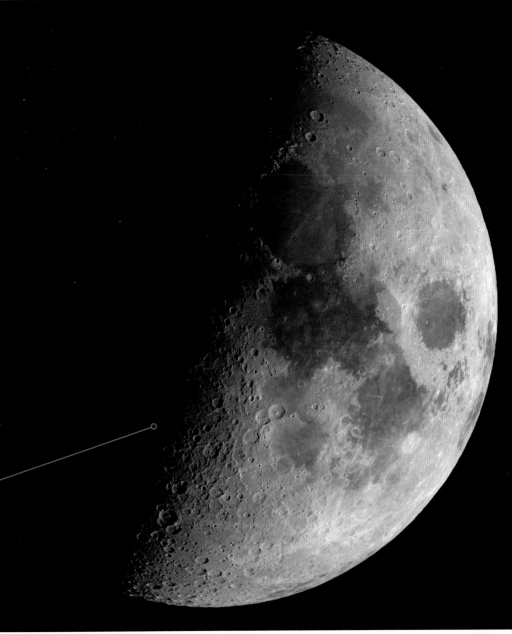

The terminator is the line that divides the bright and dark sections of the Moon. It's the best place to look for details.

NASA

Cut down the moonlight

Many observers use either neutral density filters or variable polarizing filters to reduce the Moon's light. I prefer to use the latter because an observer can change how much light a variable filter transmits.

Two other methods to reduce the Moon's brightness are to use high magnification or to add an aperture mask. High powers restrict the field of view, thereby reducing light throughput. An aperture mask causes your telescope to act like a smaller instrument while maintaining its focal length.

Use your best vision

Turn on a white light behind you when you observe between Quarter and Full phases. The light should be moderately bright (something in the 60-watt range), but neither your eyes nor the eyepiece should be in direct view of the fixture. The addition of white light suppresses the eyes' tendency to dark adapt at night. Not dark adapting causes the eye to use normal daytime vision, which is of much higher quality than dark-adapted night vision. In essence, you'll see more detail because you're viewing with a better part of your eye.

Work from a list

One of the best ways to learn the Moon is to undertake an observing project. In the United States, the Astronomical League offers one such project: the Lunar Observing Program. You'll learn a lot about our satellite as you work through a list of 100 lunar features. To receive a certificate, you must be a member of the league, either individually or through an astronomy club. For details, go to www.astroleague.org and choose "Observing Programs" from the dropdown menu under "Observe" at the top of the page. In the U.K., the British Astronomical Association coordinates lunar observing. Visit www.britastro.org and choose "Lunar" from the menu on the right side of the homepage.

Dig for the details

Most of the Moon's named features are craters. Challenge yourself to see either how small a crater you can detect or how many craterlets (small craters) in a given area you can observe. You'll need a detailed Moon map for this project. When looking for craterlets, you can search a lunar mare (sea), but usually a large, flat-bottomed crater works best. For example, within the large crater Plato, you'll find four craterlets, each about 1.2 miles (2 km) across. Lunar observers consider seeing these craters a test for a 6-inch telescope.

Shoot the Moon

How can a celestial object that's so easy to photograph be so difficult to photograph well? The Moon is large and bright (only the Sun outshines it), and you can use any camera connected to any size telescope to image it. That's the easy part. But the Moon also contains vast areas of low contrast that have little color differential. Recording those regions so they look like what your eyes see is the hard part. Luckily, we live in the digital age. Unlike when astrophotographers used film, it now costs nothing extra to take 200 images instead of just one. Examine them, throw away what you don't like, change one or more parameters (including the techniques you use to process the images), and shoot more as you continue to perfect your techniques.

▶ A NEUTRAL DENSITY filter reduces the amount of light that strikes your eye, making bright objects like the Moon easier to observe. MICHAEL E. BAKICH

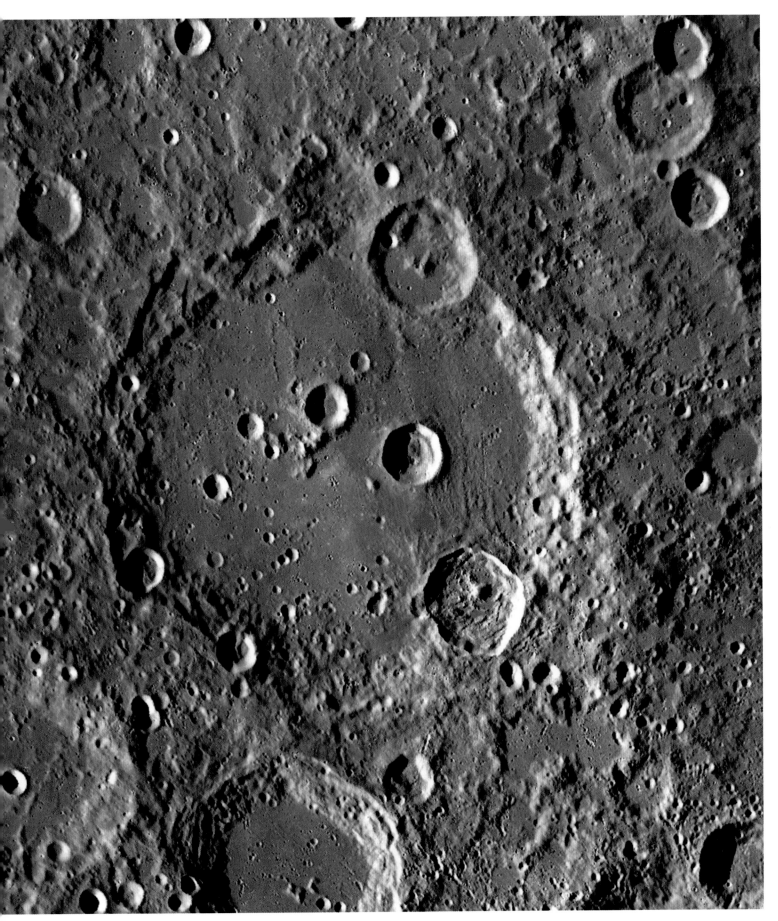

Observing Mercury

Five planets are visible to the naked eye. Of these, Mercury is the most difficult to observe. Indeed, many amateur astronomers have never seen Mercury.

Mercury orbits the Sun at an average distance of only 36 million miles (58 million km). Because Earth is nearly three times as far away, from our perspective, Mercury always stays near the Sun.

When Mercury is as far east of the Sun as it can get (called greatest eastern elongation), we see it as an evening star low in the west. When it's west of the Sun, we view it as a morning star in the east before sunrise. Some elongations are better than others because of Earth's tilt and the stretched-out nature of Mercury's orbit. Even at its farthest from the Sun, Mercury appears no more than 28° away. At a "bad" elongation, the planet may be as little as 18° from the Sun.

But the numbers above only give angular distance from the Sun. For observers to see a good elongation of Mercury, the Sun's apparent path through our sky — called the ecliptic — has to be nearly at a right angle to the horizon. This position puts Mercury higher in the sky than when the ecliptic/horizon angle is steep.

Through a telescope, Mercury disappoints. First, its disk appears small: only 7" across at greatest elongation. Second, it never appears high in the sky. So, when you observe Mercury, you're looking through the thickest, most distorting part of Earth's atmosphere.

You can address this last problem in a way that may sound strange to beginning amateur astronomers: Observe Mercury during the day, when it appears highest in the sky. Remember to use extreme caution when attempting this — the planet never strays far from the Sun. A go-to drive or a manual telescope drive that you previously aligned and left on after sunrise works well for this observation. (For advanced observers, if your telescope's mount has setting circles, use them to find Mercury's position by offsetting from the Sun's coordinates.) You can improve your daytime telescopic view of the planet if you use a yellow, orange, or red filter to reduce the amount of scattered blue light. Through your scope, expect to see Mercury go through phases similar to the Moon's.

Most observers will not be able to detect surface markings on Mercury. It takes a seasoned observer and excellent atmospheric conditions to see anything at all on the planet, even through the largest amateur telescopes. Experienced amateurs, however, have recorded dusky markings and occasional bright areas on the planet.

For most observers, however, just seeing Mercury in the evening sky counts as a successful observation.

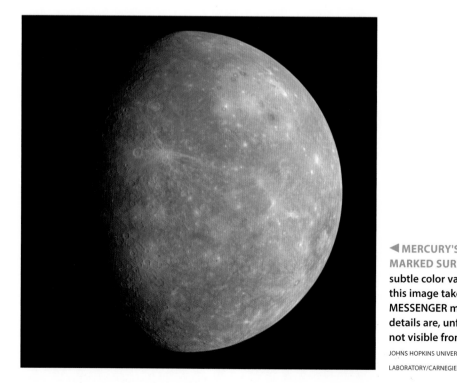

◀ MERCURY'S POCK-MARKED SURFACE shows subtle color variations in this image taken by NASA's MESSENGER mission. Such details are, unfortunately, not visible from Earth. NASA/ JOHNS HOPKINS UNIVERSITY APPLIED PHYSICS LABORATORY/CARNEGIE

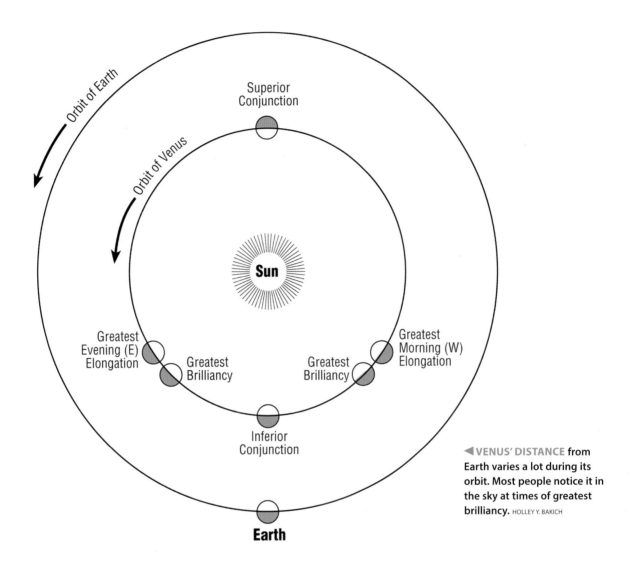

Superior
Conjunction

Orbit of Earth

Orbit of Venus

Sun

Greatest
Evening (E)
Elongation

Greatest
Brilliancy

Greatest
Brilliancy

Greatest
Morning (W)
Elongation

Inferior
Conjunction

Earth

◄ VENUS' DISTANCE from
Earth varies a lot during its
orbit. Most people notice it in
the sky at times of greatest
brilliancy. HOLLEY Y. BAKICH

Observing Venus

Because it's so bright and its appearances in the morning and evening skies last for months, Venus is easy to observe. During its orbit, the planet goes through a pattern of inferior conjunction (when it's between Earth and the Sun), greatest western elongation (when it's at its maximum distance from the Sun in the morning sky), superior conjunction (when it's on the other side of the Sun from Earth), and greatest eastern elongation (when it's at its maximum distance from the Sun in the evening sky).

One more event — greatest brilliancy — occurs approximately 36 days before and after inferior conjunction, when the planet lies 39° from the Sun. The geometry of the Sun-Venus-Earth angle at these times makes Venus appear brightest from Earth.

Like Mercury, Venus undergoes phases. The phases of Venus are of interest to observers, as is another easy-to-see aspect: size change. Mercury looks twice as big near inferior conjunction as it does at superior conjunction. Venus, on the other hand, is more than six times larger.

Daytime observations of Venus are not as difficult as most amateur astronomers imagine. In fact, it's far better to observe Venus during the daytime, or at least in twilight, because the background sky brightness reduces the deleterious effects of the planet's brilliance. And observing Venus in the daytime sky is easy:

Simply point your telescope at Venus before daybreak and allow the drive to track it until after sunrise.

The problem with daytime observations is that solar heating of the ambient air (and your telescope) can produce some really bad seeing (atmospheric steadiness). Most locations report the worst daytime seeing in the afternoon, so aiming to observe the planet before noon is generally best.

If you observe Venus at night, limit your viewing to when the planet is at least 20° above the horizon. The air below that level is so thick that image quality will suffer.

Observing Mars

Observing the Red Planet was once described to me by a friend as "two long years of waiting for four to six weeks of panicked activity." Although Mars looks best during its brief opposition, it exhibits a wide range of sizes. Don't wait for it to reach opposition before you start observing it. Quality observations can be made even when the planet's apparent diameter is under 10". As long as the atmosphere of Mars is transparent, you'll see detail.

Another reason not to wait for Mars to reach maximum size is when opposition occurs near martian perihelion (when the planet lies closest to the Sun). Sometimes dust storms, caused by solar heating, have raged over huge areas of the planet. If you wait until opposition, there's a chance you won't see much. When opposition occurs near aphelion (Mars farthest from the Sun), the atmosphere is essentially dust-free and the features stand out prominently.

The rotational rate of Mars is 37.4 minutes longer than Earth (which takes 1,440 minutes, or 24 hours, per rotation). If you were to observe Mars *at the same time* each day, its markings will appear to gradually change by $(37.4/1,440) * 360° = 9.35°$ per day to the west. In a little more than five weeks, the planet would appear to slowly rotate backward. All the prominent features of Mars would, at some time during this period, be placed favorably on its meridian. You can also wait for Mars' rotation to bring an object into view or onto its meridian. Since Mars rotates once every 24.623 hours, in one hour it will rotate $360°/24.623 = 14.62°$.

Use high magnification to observe Mars. There's a lot of small detail on the planet, so give yourself the best chance to spot it. Prepare to spend a lot of time at the eyepiece, waiting for moments of good seeing. When they arrive,

▲ **AMATEUR TELESCOPES** reveal far fewer details in the clouds of Venus than is shown in this image taken by the Pioneer Venus Orbiter. NASA

Amateur astronomers have reported seeing an irregular terminator, dusty shadings, bright spots, and caps on the cusps, to name the most obvious sightings. Viewed in visible light, there are no permanent features discernible in the clouds of Venus. The atmosphere is in a continuous state of mixing, and any patterns observed quickly dissipate.

The best — really, the only — way to see features in the atmosphere of Venus is through a dark blue (No. 38A) or violet (No. 47) eyepiece filter. Such filters, unfortunately, don't allow much light through. They have transmissions of 17 percent and 3 percent, respectively. So this advice really is for those who have access to a 10-inch or larger telescope.

The most-reported sighting using these filters is of an immense C- or Y-shaped feature centered on and symmetrical with the planet's equator. This is a short-lived phenomenon, but it tends to re-form often enough to be considered a "permanent" feature in the clouds of Venus.

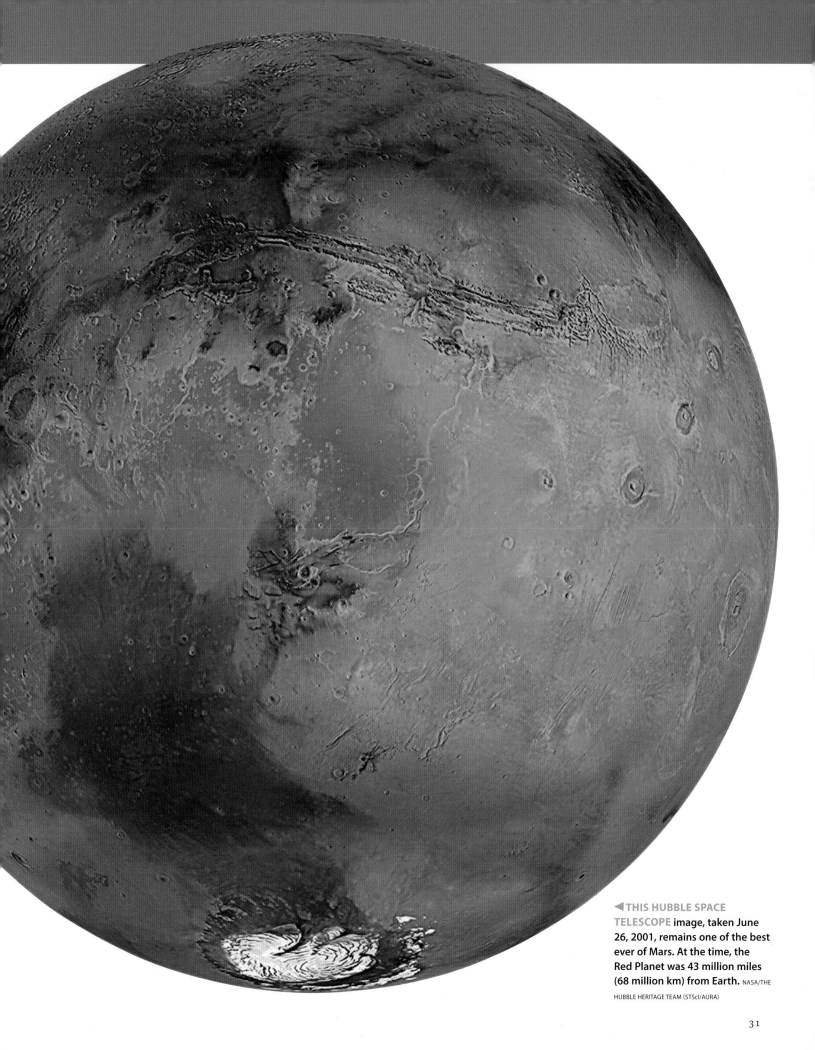

◀THIS HUBBLE SPACE
TELESCOPE image, taken June
26, 2001, remains one of the best
ever of Mars. At the time, the
Red Planet was 43 million miles
(68 million km) from Earth. NASA/THE

HUBBLE HERITAGE TEAM (STScI/AURA)

DISCRETE CLOUDS
are generally related to one area and are carried along as the planet rotates. Most are found in Mars' northern hemisphere during the spring and summer.

▲ **THESE NIGHT-GLOWING clouds on Mars were spotted May 18, 2019, by the Curiosity rover.** NASA/JPL-CALTECH/KEVIN M. GILL

concentrate your attention on a small area or a single feature. Trying to see the entire disk all at once or darting your gaze from one feature to another is not a profitable use of your observing time. Compare your views to a detailed map of Mars.

Dust storms are more likely to occur near martian perihelion, but they can develop anytime. Summertime (on Mars) dust storms are generally larger and have greater coverage. They can either be localized dust storms, associated with a desert region, or global dust storms. True global dust storms were not observed until 1956. Then six were seen in a span of less than 50 years.

Several different types of clouds are observable in the martian atmosphere. Some are *seasonal* clouds and are related to the seasonal heating and cooling, which causes sublimation and condensation of water and carbon dioxide ice.

"Discrete" clouds are generally related to one area and are carried along as the planet rotates. Most are found in Mars' northern hemisphere during the spring and summer.

Certain discrete clouds are known as "orographic" clouds. These are caused by wind passing over the high peaks of mountains and volcanoes and are composed of water. To view the high-altitude orographic clouds, use a blue or violet filter. For the low-altitude ones, a green filter works better.

A good medium-size telescope observing challenge would be to try to observe the Syrtis Blue Cloud. This is a famous discrete cloud associated with the Libya basin and Syrtis Major.

Because this cloud turns Syrtis Major a bluish color, use a yellow filter and the part of Syrtis Major covered by the cloud will appear greenish.

Morning and evening clouds may also be observed. These are bright, isolated patches of surface fog seen at sunrise (the western edge of Mars) or sunset. You may also spot ground frost. The difference is that the fog usually dissipates in a few hours, while the frost may last all day. Evening clouds are generally larger and there are more of them. They tend to grow as night approaches. Telescopic views of morning and evening clouds are enhanced through blue or violet filters.

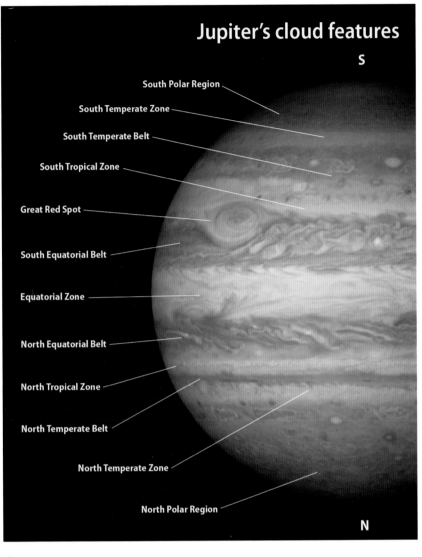

Jupiter's cloud features

S

- South Polar Region
- South Temperate Zone
- South Temperate Belt
- South Tropical Zone
- Great Red Spot
- South Equatorial Belt
- Equatorial Zone
- North Equatorial Belt
- North Tropical Zone
- North Temperate Belt
- North Temperate Zone
- North Polar Region

N

▲ **JUPITER DOESN'T** have a solid surface we can see, so observers use this nomenclature to describe its clouds.

NASA/*ASTRONOMY*: ROEN KELLY

Observing Jupiter

After the Sun and Moon, the celestial object with the greatest detail is Jupiter. Even small telescopes show the four Galilean satellites, Io, Europa, Ganymede, and Callisto. They appear as bright "stars" on either side of Jupiter.

Several dark stripes are easy to see. They lie on either side of the equator and are the North and South Equatorial Belts. Through a normal telescope (which flips the image), south is up and features move across the globe from right to left (west to east). More belts and zones are visible through larger scopes. At high magnification, see if you can notice that Jupiter is flattened, a result of its rapid rotation rate coupled with the fact that it is not a solid planet.

The planet's fast rotation brings nearly all of its visible area into view in a single night around the time of opposition. There are also times when individual belts and zones become more or less prominent. Some have even disappeared for extended periods of time.

Large telescopes and good seeing will let you see details on the moons. With powers of 350x and higher, you might resolve disks, especially during transits when the satellite's glare is less due to Jupiter's lighter background. Ganymede is the best candidate for this. Look for the lighter shaded frost in its polar regions. With the biggest amateur scopes, observers have even seen the colors of these satellites.

Many filters work well on Jupiter. Try a blue (#38A) filter to enhance the dark reddish-brown belts. A red (#23A) filter will bring out blue features like festoons in the Equatorial Zone as well as the northern and southern borders of the major belts. Red filters also brighten and enhance white spots and ovals in the South Temperate Belts and Zone.

▼ **THE JUNO SPACECRAFT** flew directly over Jupiter's south pole February 2, 2017, to acquire this image. At the time, it was 62,800 miles (101,000 kilometers) above the cloud tops. NASA

South Polar Region
South South Temperate Zone
South South Temperate Belt
South Temperate Zone
South Temperate Belt
South Tropical Zone
South Equatorial Belt
Equatorial Zone South
Equatorial Band
Equatorial Zone North
North Equatorial Belt
North Tropical Zone
North Temperate Belt
North Temperate Zone
North North Temperate Belt
North North Temperate Zone
North Polar Region

C ring
B ring
A ring
Encke Division
Cassini Division

▲ WHEN YOU OBSERVE Saturn, you'll see features on the planet and in its magnificent ring system. *ASTRONOMY: ROEN KELLY*

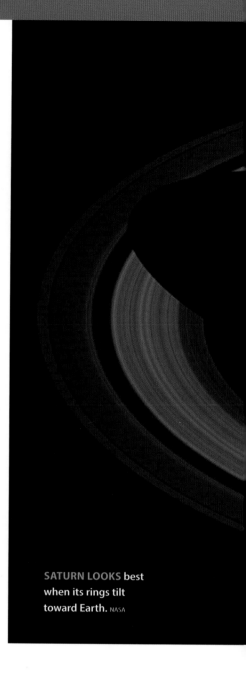

SATURN LOOKS best when its rings tilt toward Earth. NASA

Observing Saturn

The standard telescope recommendation for detailed studies of Saturn is a high-quality 6-inch scope. It will show a series of white or pale yellow zones and darker yellowish-brown or bluish-brown belts running across the globe parallel to Saturn's equator.

Saturn's globe casts a shadow on the ring system to the left (east) prior to opposition, to the right (west) after opposition, and casts no shadow at opposition.

Through small telescopes, the obvious features on the globe are the Equatorial Zone (EZ), the North and South Equatorial Belts (NEB and SEB), and the South Polar Region (SPR). Observers with large telescopes may see

the SEB divided into north and south components (SEBn and SEBs). Between them lies a brighter region: the South Equatorial Bright Zone (SEBZ). Some less distinct belts and zones, such as the South Tropical Zone (STrZ), South Temperate Belt (STeB), and South Temperate Zone (STeZ), should be visible through 10-inch and larger scopes during good seeing.

Discrete phenomena in the belts and zones, such as dusky festoons and bright ovals, also come into view occasionally. Smaller and less conspicuous white ovals may appear unexpectedly at any time.

Filtering your view

Colors on the disk of Saturn are less bright and show less contrast than those you'll see on Jupiter. With a large enough scope, you may be able to see a difference in the color of the rings. For example, one side of the rings may appear brighter than the other when you view through a red filter. If so, switch to a blue filter and see if the brightness variation reverses.

As for the colors themselves, observers describe the brighter zones as appearing off-white, slate-gray, or yellowish at times. Saturn's belts exhibit bluish-gray, brown, and reddish colors easily seen through light red (#23A), red (#25A), orange (#21), or yellow (#12) filters. Brighter patches sometimes appear

on the Ringed Planet and look best through a green (#58) filter. To highlight the rings, use a light green (#56) or a light blue (#80A) filter. Intensity estimates of belts and zones made while using color filters are valuable for monitoring the prominence of atmospheric features in different wavelengths.

The ring system

The broad rings encircling Saturn's globe are responsible for the planet's exquisite beauty and popularity. Through a medium-size telescope, it is easy to see the rings divide into three main components of varying brightness.

The outermost of these is the A ring. A dark gap called the Cassini Division, which is visible through a 3-inch telescope, separates it from the brighter central B ring. The innermost of the major ring components is the C ring, by far the faintest of the three. Just seeing it usually requires a 6-inch scope. Halfway between the outer and inner edges of the A ring lies the Encke Division, and eight-tenths of the way out from the globe in the A ring is the very narrow Keeler Gap.

Most observers agree that seeing the Encke Division requires a telescope aperture of about 8 inches, and the Keeler Gap is within reach of 10-inch instruments.

Saturn's moons

Of Saturn's many satellites, only about seven have significance to observers. Mimas, at visual magnitude 12.1, is the faintest of them. It's difficult to find because it lies close to Saturn, so seeing Mimas requires a large aperture. Enceladus, at visual magnitude 11.7, fluctuates in brightness and also is hard to detect with anything smaller than a 10-inch telescope.

Tethys and Dione are both roughly visual magnitude 10.6, so you can spot them through a 4-inch scope when they're far enough from Saturn's globe. Rhea, at visual magnitude 10.0, is easy to see using a 3-inch telescope. The Ringed Planet's brightest moon, reddish Titan, which shines at magnitude 8.3, makes an easy target through a 2.4-inch instrument.

But perhaps the most interesting saturnian satellite is Iapetus. Its visual magnitude ranges from 10.1 to 11.9, depending on where it is in its orbit around Saturn. From data returned by the Voyager spacecraft, we know that one face of Iapetus is bright, reflecting 50 percent of the Sun's light, while the other is quite dark, reflecting only 5 percent of the light that strikes it. Observers can see Iapetus through a 3-inch telescope when the moon is at its brightest.

▼ **ENCELADUS IS a moon of Saturn that astronomers think may have a subsurface ocean.** NASA

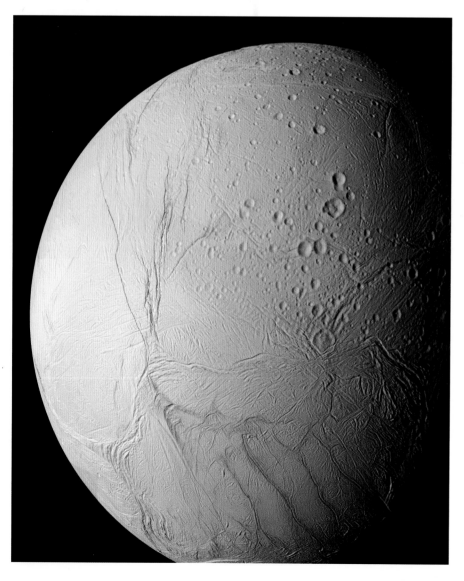

Observing Uranus and Neptune

Astronomers call these two worlds the ice giants. This term refers to their compositions and distinguishes them from the gas giants Jupiter and Saturn, which are mainly hydrogen. Uranus and Neptune contain much less hydrogen and feature other elements, such as carbon, nitrogen, oxygen, and sulfur as well as proportionally larger icy, rocky cores.

Uranus, which lies an average of 1.8 billion miles (2.9 billion kilometers) from the Sun, and Neptune, 2.8 billion miles (4.5 billion km) away, are on every amateur astronomer's "life list." Just seeing the ice giants through binoculars or a small telescope is a score for beginners. But what can you see if you look carefully and through a medium or large scope?

It's best to observe Uranus and Neptune near

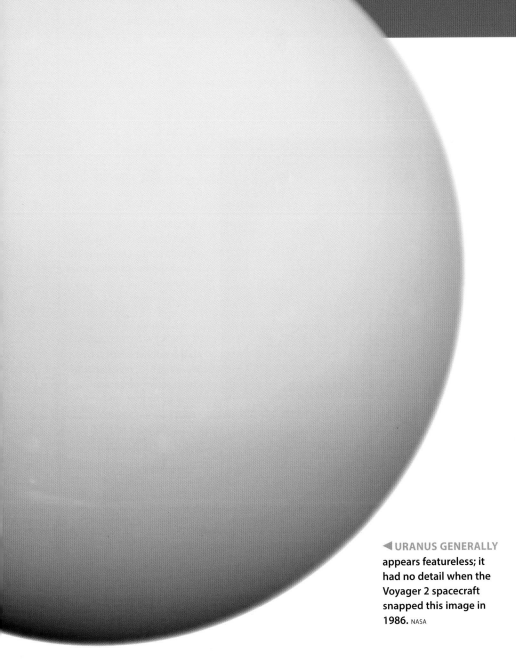

— and Neptune's largest, moon, Triton. The first two glow at magnitudes 13.2 and 13.4, respectively, when Uranus comes to opposition. Triton shines at magnitude 13.0 when Neptune reaches its peak, and doesn't rapidly deviate from that value. To see any of these moons, you'll need a 10-inch or larger telescope and an eyepiece that magnifies at least 150x. And to answer your question, yes, 200x or 300x would be even better.

These moons' faintness is only part of the problem, however: None strays far from its host. Oberon never appears more than 44" from Uranus (about 12 planetary diameters away) while Titania maxes out at 33" (nine times Uranus' diameter). Triton is the easiest of the three to see, even though it always remains within 17" of Neptune (not quite seven and a half Neptune diameters). It leads because it glows slightly (20 percent) brighter than the others and Neptune produces much less glare than Uranus.

The key to finding any of these moons is knowing how, when, and where to look. First, choose a dark observing site with good seeing. Turbulence in Earth's atmosphere will cause planetary images to appear mushy and wash out these faint moons entirely. You'll also want to avoid glare from Earth's Moon. Try to observe within three or four days of a New Moon.

◀ **URANUS GENERALLY appears featureless; it had no detail when the Voyager 2 spacecraft snapped this image in 1986.** NASA

▼ **NEPTUNE LOOKS blue and tiny through telescopes, but the Voyager 2 spacecraft revealed some details in its cloud tops.** NASA

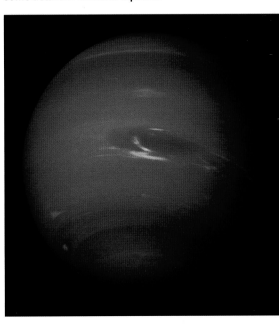

opposition. The orbit of Uranus tilts less than 1° from the plane of the ecliptic, so you'll always find it close to that line. Uranus' average apparent motion (against the background stars) is approximately 42" per day. It takes Uranus about 44 days to move the width of the Full Moon. Neptune, with an average apparent motion against the background stars of only 22" per day, takes approximately 85 days to traverse a distance equal to the width of the Full Moon.

The visible atmosphere of Uranus is generally a featureless haze. But throughout the planet's history, many observers have seen details. The first such observer, British astronomer William Buffham, noticed two round bright spots and a bright zone in 1870. In 1883, American astronomer Charles Augustus Young reported markings, along with both polar and equatorial belts. These details, unfortunately, have one thing in common: They were exceedingly faint. Even the Voyager 2 mission that flew past Uranus in January 1986 showed a nearly featureless globe.

A telescope won't reveal much in the way of surface features on Neptune's tiny disk. Basically, an amateur astronomer can't do much better than identifying the planet among countless background stars. The thrill to observing Neptune is simply finding it.

Merely spotting the ice giants may be enough satisfaction for some novice observers, but if you're in the mood for something a bit more difficult, why not try to spot one of these worlds' satellites? Plan to spend some serious time on this task. Although the planets are relatively easy to find, their moons are quite hard to pick out from the starry background. Most amateur astronomers have a realistic hope of spotting two satellites of Uranus — Titania and Oberon

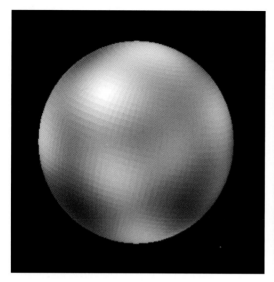

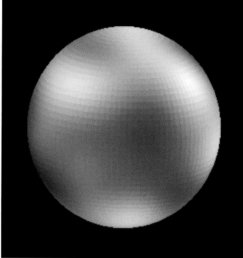

◀ **PLUTO REVEALS** little detail from Earth, even to the Hubble Space Telescope. This image shows opposing sides of the world. A. STERN (SwRI)/M. BUIE (LOWELL OBS.)/NASA/ESA

▼ **THE NEW HORIZONS** spacecraft, which flew by Pluto in 2015, revealed it to be an active world that requires further study. NASA/JHUAPL/SwRI

Observing Pluto

As seen from the Earth, Pluto moves slowly. Its average apparent motion against the background of stars is a leisurely 14" per day. It takes Pluto roughly 130 days to travel a distance equal to the width of the Full Moon.

You can see Pluto through an 8-inch telescope, although you will need a dark site and an accurate star chart to do so. The easiest way to observe Pluto is on successive nights. Use a star chart that shows Pluto and stars down to magnitude 15. Remember to invert and/or reverse the printed image to match what you see through your telescope.

At the eyepiece, identify the position of any object in the expected location of Pluto as shown on the star chart. The following night (or the next clear night) find the same star field and search for Pluto. After noting the exact date and time, mark what you believe to be the new position of Pluto on your original map. Finally, find the position of Pluto that corresponds to the new date and time and compare. If the position given by your software matches the position of the mark on the original star chart, you have successfully observed Pluto.

COMET HYAKUTAKE (C/1996 B2) passed closest to Earth on March 25, 1996.

NASA/BILL INGALLS

Observing asteroids

Although asteroids are essentially point sources and appear as stars in amateur telescopes, many are bright enough to be observed even through instruments as small as a 4-inch refractor. The key to recognizing an asteroid visually is to observe it several times. One method is to make a sketch of the star field at two different times. The second method is to mark a printed star chart with the position of the possible asteroid and then return to the field at least an hour later. Observe asteroids visually near opposition when they are brightest and when their relative motion is greatest.

Observing comets

If the comet is bright enough, start with a detailed naked-eye observation from a dark site. Try to determine the full extent of the tail: How wide is it, both near the coma and at the tip? Can you see both a dust tail and a gas tail? How do they differ? Finally, note the color of all parts of the comet, especially the tail. If you choose to sketch the comet, be sure to note the direction of north on your sketch.

Next, observe the comet through binoculars. Now how far from the comet's head can you observe the last wisps of the tail? Also, examine

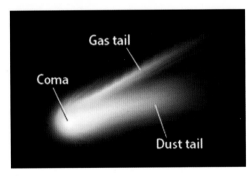

▲ COMETS TYPICALLY have two tails.

ASTRONOMY: ROEN KELLY

the width of the comet, especially the tail. If both tails are visible, note any increased definition in either and any colors seen. Also note the separation and shapes of the two tails.

Finally, examine the comet through your telescope. Use a variety of magnifications and take your time. Note any irregularity in the shape or brightness of the coma. Does it look elliptical (and to what degree) at all magnifications? The nucleus of any comet is too small to resolve, however you may see the *pseudonucleus* as a condensed or starlike area. Does the brightness of the coma vary with distance from it, and if so, how?

On a number of occasions, even observers with medium-sized telescopes have seen fragmenting within the coma. If you're lucky enough to see this, the comet will appear to have

several pseudonuclei. When you observe at high power, also be aware of the possibility that you may see jets. These features will appear as lines or angular rays, generally in the direction of the Sun because solar heating on the sunward side of the comet creates outgassing.

Estimating a comet's brightness is not a simple thing to do. One of the mistakes amateur astronomers make is to underestimate the total brightness of the coma. They concentrate too much on the starlike central portion. Also, from a less-than-dark site, the outer coma is often lost to sight.

One method involves defocusing the images of nearby stars to the same size as the normal, in-focus view of the comet's coma and comparing brightnesses. This technique works best for diffuse comets.

Another way involves defocusing the images of the comet and surrounding stars to such an extent that the circles can be directly compared. Some observers take this a step further by continuing to defocus the images until objects start to disappear. By noting the brightnesses of objects that disappear, especially ones that disappear just before and just after the comet itself, one can roughly estimate the brightness of the comet.

SHOWER METEORS appear faster the larger their distance from the radiant (up to 90°) and the higher their elevation above the horizon.

Observing meteor showers

Many observers have shared tips related to meteor shower observing with me. Here are a few I consider the most beneficial.

• Look about two-thirds of the way from the horizon to the zenith. Don't place the radiant in the center of your field of view. That's where the shortest trails will appear.

• Pick a night near New Moon and observe from the darkest site possible. Mark the position of the radiant on a star chart and, when darkness falls at the site, find that position in the sky and memorize it.

• If you observe in a group, encourage every observer has to observe independently. Avoid

combining data from different observers. Each observer should keep their own notes and fill in their own report forms.

• Always note the magnitude of the faintest star you can see near the zenith. The limiting magnitude is a quantity different for each observer. It not only defines the sky conditions but also the quality of the observer's eyes.

• Shower meteors appear faster the larger their distance from the radiant (up to 90°) and the higher their elevation above the horizon. Near the radiant or near the horizon, shower

meteors generally appear to move slower.

• Avoid talking and listening to music, as it might distract others. When you get tired, take a break to walk around or eat something.

• Most observers follow the motion of the sky, and thus become more familiar with certain regions. If your chosen field becomes cloudy or too low, choose another. Be sure to record the center of the new field in your notes.

• Note your observations on an audio recorder or on paper. Don't take your eyes off the sky. (This takes practice if you are using paper.) Your notes should include 1) the time of

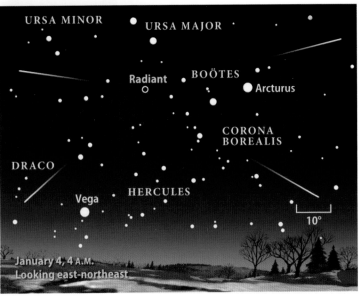

◀ THIS BRILLIANT meteor from the Perseid meteor shower lit up the sky August 12, 2016. WIKIMEDIA COMMONS

▼ OBSERVERS CAN trace all meteors belonging to a meteor shower back to its radiant. This illustration shows the radiant of the Quadrantid meteor shower, which peaks in early January.

ASTRONOMY: ROEN KELLY

the beginning and end of the meteor watch with all breaks noted; 2) the limiting magnitude and any changes to it during your watch; 3) the details of any cloud cover; 4) the right ascension and declination of the center of the field of view; and 5) the details of all meteors observed, such as magnitude, location (was it a shower meteor or sporadic?), color, and the presence or absence of a trail.

• Do not look at specific constellations. Rather, try to see only a random collection of stars. More difficult than this, try not to concentrate on the stars at all. You want to watch the larger field, not the points of light.

OBSERVING THE DEEP SKY

Observing double stars

Astronomers estimate that 60 percent of all stars are double or multiple stars. Observing double stars is fun, easy, and rewarding. It doesn't take a complicated setup, you can observe from a city, and stellar pairs exist to challenge every size scope.

In addition to the double star's location and how bright each component is, a double-star observer needs to know two quantities. The first is the pair's apparent angular separation. This number is given in arcseconds (") and it's simply the distance between the two stars.

The second quantity is the position angle. This is the angle, measured from north through east, of the fainter star of the pair (the companion or secondary) from the brighter (the primary) star. If the companion is due north of the primary, its position angle is 0°; if it's due east, 90°; if midway between south and west, 225°. Which double stars you can split depends on the size of your telescope because the resolution of a telescope depends largely on its size.

Seeing double stars with close separations requires two things: high magnification and great atmospheric conditions. Resolving double stars of similar magnitudes is straightforward. It's more difficult when the companion differs from the primary by several magnitudes. In some cases, you'll find a brightness difference of up to 10 magnitudes in a double-star system, as with Sirius.

Most observers find colorful double stars a joy. It takes some time to train your eye to see colors through a telescope, but the payoff is big. The separation of the stars often helps in color identification. The contrast between two or more stars in close proximity brings out subtle color tones normally lost if you view each star separately.

The best double stars to observe are those with contrasting colors. Who among us can look at Albireo (Beta Cygni) and not be amazed? The contrast between stars colored gold and sapphire never fails to delight.

▲ **THIS SKETCH** of Albireo (Beta Cygni) was made using an 8-inch telescope and an eyepiece that gave a magnification of 200x. JEREMY PEREZ

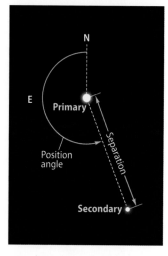

◀ **THE POSITION ANGLE** for binary stars starts at north and proceeds through east; separation is measured in arcseconds. The A star is the primary, and the B star is the secondary. HOLLEY Y. BAKICH

/// OBSERVE THESE BRIGHT NEBULAE

Designation	R.A.	Dec.	Size	Map
NGC 281	0h53m	56°36'	35' by 30'	1
NGC 1499	4h01m	36°38'	160' by 40'	5
NGC 1931	5h31m	34°12'	4' by 4'	5
M1	5h35m	22°02'	8' by 2'	5
M42	5h35m	−5°28'	90' by 60'	11
NGC 1999	5h36m	−6°44'	2' by 2'	11
IC 434	5h42m	−2°27'	90' by 30'	11
NGC 2170	6h07m	−6°23'	2' by 2'	11
NGC 2175	6h10m	20°29'	40' by 30'	5
NGC 2264	6h41m	9°54'	10' by 7'	11
M20	18h02m	−23°00'	20' by 20'	20
M8	18h04m	−24°20'	45' by 30'	20
NGC 6559	18h10m	−23°59'	15' by 10'	20
NGC 6590	18h17m	−19°44'	4' by 3'	20
M16	18h19m	−13°49'	120' by 25'	14
M17	18h21m	−15°59'	40' by 30'	14
NGC 6888	20h13m	38°19'	20' by 10'	9
NGC 6914	20h25m	42°23'	3' by 3'	9

R.A. = Right ascension; Dec. = Declination

/// THE DARKEST DARK NEBULAE

Designation	R.A.	Dec.	Area	Map
LDN 1506	4h20m	25°17'	0.334	5
LDN 1535	4h36m	23°54'	0.111	5
LDN 1544	5h04m	25°14'	0.109	5
LDN 1622	5h55m	2°00'	0.122	11
LDN 1709	16h33m	−23°46'	0.099	20
LDN 204	16h48m	−12°05'	0.167	14
LDN 162	16h49m	−14°15'	0.124	14
LDN 65	17h13m	−21°54'	0.088	20
LDN 513	18h11m	−1°33'	0.127	14
LDN 557	18h39m	−1°47'	0.181	14
LDN 530	18h50m	−4°47'	1.124	14
LDN 673	19h21m	11°16'	0.199	14
LDN 694	19h41m	10°57'	0.109	14

LDN = Lynd's Dark Nebula; R.A. = Right ascension;

Dec. = Declination; Area given in square degrees

THE HORSEHEAD NEBULA (Barnard 33) probably ranks as the sky's best-known dark nebula. Several million years from now, the dust and gas in this region will become new stars. ADAM BLOCK/MOUNT LEMMON SCIENCE CENTER/UNIVERSITY OF ARIZONA

Nebula (M16, Map 14), the Swan Nebula (M17, Map 14), the Trifid Nebula (M20, Map 20), and M43 (Map 11).

Dark nebulae are clouds of dust and cold gas. We see them only because they obscure light from stars or bright nebulae behind them. Some — for example, the Horsehead Nebula (B33, located in IC 434, Map 11) in Orion — are small and difficult to see even through large telescopes. Others, such as the Coal Sack in Crux (Map 23), are large and easy to see unaided.

A planetary nebula is the most common end product of stellar evolution. Stars about as massive as the Sun eject their outer layers near the end of their lives. The ejected gas becomes a planetary nebula — a spherical shell of thin matter expanding into space at more than 20,000 mph (32,187 kph). The central star emits lots of ultraviolet radiation, which causes the gas of the expanding planetary nebula to glow.

Roughly 1,000 planetary nebulae exist in our neighborhood of the galaxy. A typical one measures less than 1 light-year across. What color you see depends on your eyes. The dominant color is close to the border between the eye's perceived green and blue colors. So, an observer's color response could cause the nebula's color to appear either more green or more blue.

A planetary nebula's altitude in the sky also affects its perceived color, although not to a large extent. Reddening caused by our atmosphere means any planetary viewed at a higher altitude

▲ **THE LITTLE DUMBBELL** Nebula
(M76) is a planetary nebula in
Perseus. It looks bright and colorful
in this image but is somewhat
difficult to observe. ADAM BLOCK/MOUNT
LEMMON SCIENCE CENTER/UNIVERSITY OF ARIZONA

▶ **THE GHOST OF JUPITER**
(NGC 3242) is a planetary nebula
named for a planet. That said, it
resembles Neptune a lot more
than Jupiter. ADAM BLOCK/MOUNT LEMMON
SCIENCE CENTER/UNIVERSITY OF ARIZONA

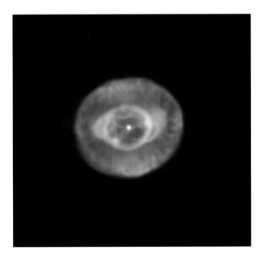

will be slightly bluer than when it's seen nearer to the horizon.

Use the biggest telescope you can when observing bright and planetary nebulae. A large scope may gather enough of the object's light to allow you to see color. Narrowband and light-pollution reduction filters also help by increasing the contrast between the nebula and the background sky.

When you observe a bright nebula, move your scope slightly to be sure you're seeing the whole object. If you can see stars, determine which are involved with the nebula. More than half of all planetaries are starlike. Most other planetaries have diameters less than 1' But some big ones exist: the Dumbbell Nebula (M27, Map 9) and the Owl Nebula (M97, Map 2). Many fine planetary nebulae have apparent diameters between 15" and 1'. Examples include the Little Dumbbell Nebula (M76, Map 1) and the Ghost of Jupiter (NGC 3242, Map 12).

Most dark nebulae are large objects best viewed through binoculars or rich-field telescopes. If you're using a scope, be sure to move it around to ensure you're seeing the entire cloud. Knowing your eyepiece's true field of view (in degrees) will allow you to estimate the size of a dark nebula.

▲ **THE DOUBLE CLUSTER** (NGC 869 and NGC 884) in Perseus is visible to sharp-eyed observers from a dark site. Binoculars really bring out its beauty.

FRED CALVERT/ADAM BLOCK/NOAO/ AURA/NSF

Observing open clusters

Open clusters lie within our galaxy's disk. We've known of a few of the brightest for thousands of years. The Beehive (M44, Map 6) in Cancer the Crab as well as the Pleiades (M45, Map 5) and the Hyades (Map 5), both in Taurus the Bull, have numerous historical references. Greek philosopher Ptolemy mentioned Melotte 111 (Map 7) in Coma Berenices and Ptolemy's Cluster (M7, Map 20) in Scorpius. Not until Galileo trained his telescope on the Beehive, however, did we learn that these objects are collections of stars. Sparse open clusters contain less than 100 stars, while rich ones may have thousands.

All open clusters are young objects, generally no more than a few hundred million years old. Few older open clusters exist because, given enough time, the stars in an open cluster disperse as they interact gravitationally with other cluster stars. If we could follow an open cluster from its formation, we'd see stars being expelled during its entire galactic journey. M67 in Cancer (Map 12) is an exception.

Astronomers have cataloged approximately 1,700 open clusters. Compared to other deep-sky objects, open clusters are large. This means you'll generally use low-power eyepieces with wide fields of view. Some clusters are large enough and bright enough that binoculars will provide more satisfying views. You can observe

hundreds of clusters through 7x50 or larger binoculars. Some open clusters also contain other celestial objects of interest. M46 in Puppis (Map 11), for example, contains a bright planetary nebula you can see through a medium-sized telescope.

Take your time when observing an open cluster. Examine the field of view closely. Try to discern stars that are members of the cluster as opposed to field stars. Usually this is easy, but it can be tricky when the cluster lies against the Milky Way. Also, large scopes sometimes can hinder the identification of

▲ **THE SALT-AND-PEPPER CLUSTER** (M37) is a rich open cluster in Auriga. It's visible at a dark site and is a wonder through a rich-field telescope. ADAM BLOCK/ MOUNT LEMMON SCIENCE CENTER/UNIVERSITY OF ARIZONA

◀ **GLOBULAR CLUSTER M3** is one of the largest and brightest such objects. It lies 34,000 light-years away and contains half a million stars. ADAM BLOCK/MOUNT LEMMON SCIENCE CENTER/UNIVERSITY OF ARIZONA

▼ **THE INTERGALACTIC WANDERER** (NGC 2419) is a globular cluster in Lynx. At a distance of 300,000 light-years, it's one of the farthest of the Milky Way's globulars. ADAM BLOCK/MOUNT LEMMON SCIENCE CENTER/UNIVERSITY OF ARIZONA

/// 20 BRIGHTEST OPEN CLUSTERS

	Object	Constellation	Magnitude	Map
1	Melotte 25	Taurus	0.5	5
2	Melotte 20	Perseus	1.2	4
3	M45	Taurus	1.5	4
4	Melotte 111	Coma Berenices	1.8	7
5	IC 2602	Carina	1.9	23
6	NGC 6231	Scorpius	2.6	20
7	M44	Cancer	3.1	6
8	M7	Scorpius	3.3	20
9	NGC 2362	Canis Major	3.8	17
10	NGC 2264	Monoceros	4.0	11
11	M6	Scorpius	4.2	20
12	NGC 4755	Crux	4.2	23
13	NGC 869/884	Perseus	4.4	1
14	M47	Puppis	4.4	11
15	M41	Canis Major	4.5	17
16	NGC 1981	Orion	4.6	11
17	M25	Sagittarius	4.6	20
18	M39	Cygnus	4.6	9
19	NGC 6633	Ophiuchus	4.6	14
20	NGC 2244	Monoceros	4.8	11

/// 20 BRIGHTEST GLOBULAR CLUSTERS

	Object	Constellation	Magnitude	Map
1	NGC 5139	Centaurus	3.5	19
2	NGC 104	Tucana	4.0	22
3	M22	Sagittarius	5.2	20
4	NGC 6397	Ara	5.3	24
5	NGC 6752	Pavo	5.3	24
6	M4	Scorpius	5.4	20
7	M5	Serpens	5.7	13
8	M13	Hercules	5.8	8
9	M12	Ophiuchus	6.1	14
10	NGC 2808	Carina	6.2	23
11	NGC 6541	Corona Australis	6.3	20
12	M3	Canes Venatici	6.3	7
13	M15	Pegasus	6.3	15
14	M55	Sagittarius	6.3	20
15	M62	Ophiuchus	6.4	20
16	M92	Hercules	6.5	8
17	M10	Ophiuchus	6.6	14
18	M2	Aquarius	6.6	15
19	NGC 362	Tucana	6.8	22
20	NGC 6723	Sagittarius	6.8	20

cluster stars because they make so many background stars visible that confusion ensues.

Observing globular clusters

In the deep-sky menagerie, no group of objects excites new observers more than globular clusters. Seasoned amateur astronomers consider them the most rewarding objects to observe. It's easy to see why. Many are bright enough to be seen even from urban settings. But when observed from a dark site, globulars explode with detail. Use high magnification and you'll explore a whole new level of detail: Faint fuzz resolves into individual, sparkling points of light that form intricate patterns.

When observing globular clusters, begin by concentrating on the constellations Ophiuchus, Scorpius, and Sagittarius (Maps 14 and 20). You'll discover nearly 70 globular clusters in these three constellations. That's right, more than one-third of all known Milky Way globulars are located in an area that's only 5.6 percent of the sky.

Globular clusters may seem just round at first glance. But carefully note the subtle shape of each. Some are slightly elliptical. Some seem to have "arms" that extend beyond the general concentration of stars. Ask yourself questions at the eyepiece: How concentrated is the cluster? What's the range of star brightnesses? How rich is the cluster? With medium-sized telescopes, you'll count dozens or even hundreds of stars that lie apart from the cluster's central condensation.

Other globulars worthy of your attention include the Hercules Cluster (M13, Map 8), M3 in Canes Venatici (Map 7), M15 in Pegasus (Map 15), and M2 in Aquarius (Map 15).

▶ **THE BLACK EYE GALAXY** (M64) lies some 17 million light-years away in the constellation Coma Berenices. A large band of dust obscures part of the spiral structure and gives this object its common name. ADAM BLOCK/MOUNT LEMMON SCIENCE CENTER/UNIVERSITY OF ARIZONA

Observing galaxies

You might wonder how objects comprising up to a trillion or more individual stars could be so difficult to observe. Of course, the answer is distance. Galaxies are so far away that, except for a few, they all appear small and faint. Advanced observers regard faint galaxies as a challenge.

This type of compulsive, competitive observing may not be for you. That's OK. There are lots of galaxies and we've listed the brightest throughout this atlas. Visually observing detailed spiral structure, like that seen in images, requires a large telescope — 20 inches or more. Smaller ones will show "mottling," which indicates the presence of spiral arms but does not constitute a true observation of them.

To see detail through medium-size telescopes, try these Messier galaxies: the Pinwheel Galaxy in Triangulum (M33, Map 4); the Whirlpool Galaxy (M51, Map 7) and M106 in Canes Venatici (Map 7); the Black Eye Galaxy in Coma Berenices

(M64, Map 7); M83 in Hydra (Map 19); and M101 and M108 in Ursa Major (Map 2).

Irregulars are the smallest class of galaxies, but most observers find them more interesting than ellipticals, the largest class. Most irregular galaxies are faint, but exceptions exist. The king of irregulars for northern amateur astronomers is the Cigar Galaxy in Ursa Major (M82, Map 2). In the Southern Hemisphere, by virtue of their nearness, the Large and Small Magellanic Clouds (Map 22) are the greatest galaxies to observe.

Take as long as you need to study these objects. It will be time well spent. NGC 55 in Sculptor (Map 16), NGC 4449 in Canes Venatici (Map 7), Centaurus A (NGC 5128, Map 19), and Barnard's Galaxy in Sagittarius (NGC 6822, Map 14) are other relatively bright irregular galaxies.

IRREGULARS are the smallest class of galaxies, but most observers find them more interesting than ellipticals, the largest class.

▲ **THE ANDROMEDA GALAXY** is the closest large spiral galaxy to our own. This Hubble Space Telescope shot is the sharpest view ever of our neighbor. NASA, ESA, J. DALCANTON (UNIVERSITY OF WASHINGTON, USA), B. F. WILLIAMS (UNIVERSITY OF WASHINGTON, USA), L. C. JOHNSON (UNIVERSITY OF WASHINGTON, USA), THE PHAT TEAM, AND R. GENDLER

◀ **NGC 4449 is a nearby irregular galaxy in Canes Venatici. Its rectangular shape is uncommon among galaxies.** ADAM BLOCK/ MOUNT LEMMON SCIENCE CENTER/ UNIVERSITY OF ARIZONA

OBSERVING TIPS

Amateur astronomy is about observing: Each and every time you look through an eyepiece, you make contact with a distant part of the universe. I've assembled this list of 50 observing tips — in alphabetical order — to help you get the most out of those precious moments behind the eyepiece. I've discovered some of them on my own, but others were passed on to me by wise observing buddies. Read them, use them, and add to them. If you do, you'll become a better observer.

1 Avoid eye fatigue

Take short breaks, and try one of these two eye exercises every 20 minutes or so. Lightly cup your eyes with your palms and relax for 60 seconds. Or simply look away from the eyepiece and roll your eyes up, down, around, and side to side for 20 seconds; then relax, eyes closed, for another 30 seconds.

2 Batteries

Take the batteries you know you'll need — as well as the batteries you think you won't need. Also: Carry a spare battery for your single-power finder. These units do not emit much light so it is very easy to leave them on.

3 Binoculars

When hand-holding large binoculars, once you focus them, move your hands toward the front of the binoculars and they will be easier to use. Also: Choose a mount for your binoculars that is quite a bit sturdier than you require. This will allow you to easily upgrade at some point in the future.

4 Camera focusing

Astrophotographers want their lenses focused at infinity, but newer auto-focus lenses can go past infinity when focused by hand. To resolve this problem, set the lens at infinity during the day and then lock it there with one or two wraps of tape around the barrel. Use tape that won't leave a residue (e.g., no duct tape). Manual focus lenses don't have this problem, but some astrophotographers tape them anyway. It's one less thing that can go wrong.

5 Collimation

When doing any type of collimation, only turn the screws a tiny bit. Large motions are NOT required. Collimation is much more critical as the focal ratio decreases (especially below about f/6). This is due to the increased curvature of the focal plane in these telescopes, resulting from a deeper mirror parabola. Make certain the primary mirror is centered in the rear of the telescope (aligned with the center line of the telescope tube). If it isn't, there is a increased chance that the front edge of the telescope will vignette some of the incoming light, which you don't want.

Finally, after all your mechanical collimation is done, perform a star collimation. A star collimation uses light reflected from all of the mirror surfaces plus the eyepiece and so is the ultimate guide.

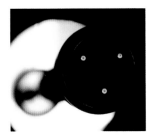

◄ **COLLIMATION** The three screws at the front of a Schmidt-Cassegrain telescope let you collimate the secondary mirror.
MICHAEL E. BAKICH

7 Condensation

If you observe on a cold night, don't immediately cover your optics when you bring your scope inside. This will allow any condensation that forms on the chilled optics to evaporate and not be trapped under a cover.

8 Dark adaption

This is the process by which the eyes increase their sensitivity to low levels of illumination. In the first 30 minutes, sensitivity increases 10,000-fold, with little gain after that. But brief exposure to bright light temporarily rolls back this hard-won increase. Just how much dark adaption you lose depends a little on the intensity and a lot on the duration of the light. A single flash from a strobe does less damage than a bright light lasting a second or more. At night, your eyes are least sensitive to red light. So, when you need to check a star chart or search for an eyepiece, use a red light, adjust its intensity to the lowest usable level, and then gaze only briefly at the illuminated object.

9 Dew shield

Dew shields installed at the front of telescopes are a good idea but will not completely stop dew from forming under harsh conditions. Consider obtaining a battery-powered dew removal system.

6 Computer light

When using a computer at an observing site, bring a piece of red cellophane or thin red plastic to cover its screen. This will help to preserve your dark adaption. The same goes for cellphones or flashlights: Use night or dark mode and opt for a red flashlight.

Red lights are by far the best color to use when observing, but make sure yours isn't too bright.
MICHAEL E. BAKICH

10 Double star trick

Try your OIII filter on difficult double stars. Some observers have reported good results on pairs of unequal brightness.

11 Eye patch

Wear an eye patch over your observing eye while setting up equipment. Put it on as long as possible before the start of your session and you will be rewarded with a fully dark-adapted eye when you're ready to begin observing. Move the patch to your non-observing eye when you look through the eyepiece. This lets you keep both eyes open, a technique that reduces eye fatigue. (Skeptical? Try reading this book with one eye closed for 60 seconds.) Before checking charts with a light, move the patch back over your observing eye.

▲ **DOUBLE STAR TRICK** An Oxygen-III filter works well on nebulae, and in some cases it may help you observe double stars. MICHAEL E. BAKICH

◄ **EYE PATCH** An eyepatch will preserve the dark adaption in your primary (observing) eye. Put it on right after sunset, and wear it until night falls. MICHAEL E. BAKICH

▲ **EYEPIECES** It's best to store eyepieces in a foam-lined case. MICHAEL E. BAKICH

▲ **FINDER SCOPE** A finder scope is a critical part of any telescope system. MICHAEL E. BAKICH

12 Eyepieces

Whether or not you store your eyepieces in a foam-lined case, always keep the plastic covers on them.

13 Finder scope

Select a finder scope with a mount that has two support rings and six adjustment screws, rather than a mount with one ring and three adjustment screws. Also: Align your finder during the day.

14 First look

Start with a low-power eyepiece in the telescope to provide a wide starting field of view.

16 Galaxies

When observing galaxies, especially in rich areas such as those in Coma Berenices and Virgo, check the field of view carefully. Your telescope may be able to reveal other galaxies not on your star map.

17 Going deep

When you're observing objects at the limit of vision or looking for small details in brighter ones, use a technique sometimes called "rocking the scope." Gently tap the mount or the telescope tube. It really helps faint details pop out!

18 High-altitude observing

This results in a degree of hypoxia, or low oxygen in body tissues, that significantly alters low-light color perception. Most people notice visual changes when they travel to altitudes approximately 2,000 feet higher than where they live, although those living at or near sea level may not begin to notice such effects until they reach an altitude of 4,000 feet.

19 Intoxication

Don't drink and drive the telescope if you're looking to do some serious observing. Why? Alcohol impairs vision.

20 Just in case

Pack a "space blanket" with your equipment. Made from a metal-coated plastic film, it will trap body heat when wrapped around you, giving you an edge in all but the most severe weather conditions. It also shields against wind and rain, weighs just a few ounces, and only costs a few dollars. Often advertised as a "survival blanket," that's exactly what it may be for you.

15 Focus

Focus each time you put your eye to eyepiece and any time you have a question about the sharpness of an image.

Be sure you know where the focuser is before you begin using any telescope. MICHAEL E. BAKICH

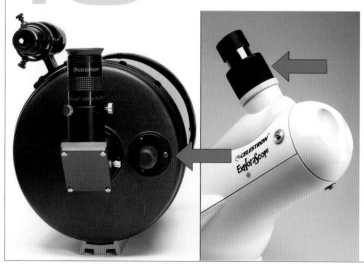

21 Know your equipment

If you've added a new piece of equipment to your observing lineup, set it up at home first. Any problem revealed in the light will be one less you'll have to deal with in the dark. As a second step, set up in your yard and observe as if you were at your remote dark-sky site. It's surprising what a simple test-run like this will teach you.

22 Limiting magnitude

No better gauge of observing-site quality exists than a direct measurement of limiting visual magnitude, or LM. Most observers determine their site's LM by identifying the faintest star they can see, usually near the zenith. Others use a method devised by meteor observers, who count the number of visible stars within predetermined asterisms. If you perform a telescopic limiting magnitude estimate, note in your observing log the telescope aperture and the eyepiece (magnification) through which the estimate was made.

23 Magnification and galaxies

Use whatever power eyepiece you must to locate the galaxy you want to observe. Then start to increase the magnification. One of the biggest mistakes that new observers make is believing that a low power will make a galaxy easier to see. In fact, using higher power will increase the contrast between the galaxy and the sky background, making the galaxy really pop out.

24 Mars

Dust storms become more active after perihelion (when the solar heating of Mars is greatest). Get in as much observing and imaging as you can prior to the Red Planet reaching this point in its orbit.

25 Mirror recoating

If you are sending your primary mirror for re-aluminizing, send your secondary mirror as well.

26 Mosquitoes

I hate mosquitoes, and I'm willing to bet you do, too. Mosquitoes attack whomever is handiest, but they prefer adults to children, women to men, and pregnant women most. They're attracted to heat, carbon dioxide, and movement, so swatting at one is essentially an invitation to others. The most effective mosquito repellents contain DEET, the acronym for N, N-Diethyl-meta-toluamide. Experts suggest treating clothing as well as exposed skin. Most fabrics are approximately 1 mm thick, but the average mosquito's proboscis is 2 mm long. DEET won't keep a mosquito from approaching, but it can stop her from biting — the attacker *is* female — by jamming receptor cells on the insect's antennae, which are sensitive to carbon dioxide and lactic acid.

27 Naked-eye deep sky

Some observers enjoy trying to spot deep-sky objects without using a telescope or binoculars. Indeed, many open and globular clusters, some nebulae, and a few galaxies are in reach of sharp-eyed observers from a dark site. If you're having trouble spotting one you think should be visible to you, locate it through binoculars first, and then go for it naked-eye.

28 Nighttime safety

If you must observe alone at a remote location, double-check everything before you leave — especially non-observing-related details, such as how much fuel is in your vehicle. Let someone know exactly where you will be and exactly how long you plan to be out, and then stick to the plan. Your life may depend on someone knowing where you are.

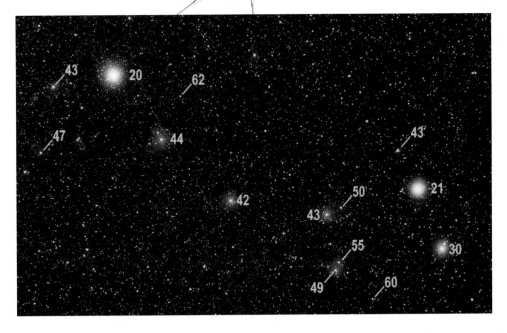

► **MAGNIFICATION AND GALAXIES** This chart shows the magnitudes of stars in and around the Little Dipper in Ursa Minor. Observers use charts like this to find a site's limiting magnitude. Note that no decimal points were used because they look like stars. HOLLEY Y. BAKICH

30

Observing chairs and ladders

If you're at all uncomfortable at the telescope, you'll do less observing and the observations you make will be less fulfilling and less accurate. In my opinion, nothing says comfort like a high-quality observing chair. Such an accessory has four main features: sturdy construction, a padded seat, easily adjustable height, and back support. Observing chairs work fine for refractors and Schmidt-Cassegrain telescopes, which have eyepieces at the lower ends of their tubes. For large Dobsonian-mounted scopes, however, a ladder of some type is usually necessary. I suggest buying a three-step folding utility ladder with a tray. Such ladders usually feature wide, rubber-coated steps — a great safety feature after sunset — and the tray is a real bonus when you need to change eyepieces or filters.

For anything other than a quick look at the sky, an observing chair is a must. An adjustable one like this can serve different size observers. MICHAEL E. BAKICH

29 Observing case
For larger objects at least, removing one fewer of the pluck-foam cubes than necessary provides a tighter fit.

31 Occulting bar
Don't wait for darkness to check your occulting bar. The best check is done in daylight. This will show any irregularities or places with greater transmission of light.

32 Polar alignment
When polar aligning your telescope, be certain to move the entire mount, not just the optical tube assembly.

33 Position of telescope
When observing sessions go well into the morning hours, be sure to point your telescope where the rising Sun won't shine. This is a crucial tip if you're going to leave your scope uncovered, but remember that the Sun has been known to damage even telescopes that were capped.

34 Question yourself
Write a few questions on index cards before your observing session. For example, a card for planetary nebulae might have these questions on it: Can you see the central star? At what magnification? What shape is the nebula? Is any color apparent? And so on. Questions will jog your memory and remind you to look for certain common details — especially when you're tired and not necessarily at your peak.

OCCULTING BAR The simplest occulting bar is a piece of tape inside an eyepiece. Make sure you get it straight. MICHAEL E. BAKICH

Observing session questions
Can you see the central star?
At what magnification?
What shape is the nebula?
Is any color apparent?

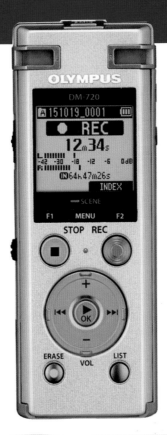

36 Site selection

Perhaps paradoxically, the finest skies on Earth are not the darkest. The more stars you're able to see, the brighter the overall sky appears. I have seen the Milky Way cast a shadow in truly dark locations. A site rated "excellent" has three characteristics: it's free of light pollution; it contains low amounts of aerosols (dust, air pollution, and water droplets); and it's at a relatively high altitude, between 5,000 and 8,000 feet (1.5 and 2.5 km).

37 Size matters

You can be the greatest observer this planet has ever seen — but if you're trying to observe extended nebulae, faint clusters, or galaxies with a 4-inch telescope, your observing log will be filled with reports of rough shapes and descriptors like "hinted at" and "small and faint." There's no way around it. If you want to observe these objects and truly get something out of the time you put in at the eyepiece, you are going to need a large telescope.

38 Sketching

Light touch-up to fill in shaded areas or to add color after your sketching session ends is permissible as long as you don't add any detail you didn't see at the telescope.

39 The sky will decide

Whether or not your telescope will provide superb views on any given night is a factor totally out of your control, no matter how well you maintain and collimate it. For planets, double stars, or anything else, the seeing — the measure of the atmosphere's steadiness — is what sets the limit on how much detail is available to your scope.

35 Record observations

I have maintained a detailed observing log for more than two decades. My instrument of choice is a digital (formerly tape) recorder. Many of my observing friends write out their observations either in a logbook or directly onto a star chart, a method that allows for sketches as well. I speak into my recorder at the telescope and transcribe the audio later, usually the next day. If you follow this path, don't let your recordings pile up.

40 Solar observing

After all precautions have been taken at the telescope, for lengthy sessions remember to apply sunscreen.

41 Star diagonal

Choose a star diagonal that will accommodate both 2" and 1¼" eyepieces (with a slide-in adapter).

42 Tripods

When choosing a tripod for your binoculars or a camera, be sure it will point to the zenith.

43 Tube currents

These degrade telescopic images, but how do you know if you have a problem? Check the out-of-focus image of a fairly bright star. If you see lots of circular motion inside the star's image, you have a severe problem. The best solution is a small, low-flow fan to move warmer air out of the telescope tube and quickly bring your mirror to the same temperature as the ambient air.

Observing nebulae

The word *nebula* is Latin for "cloud." So, a nebula is a cloud of gas and dust in space. Several types of nebulae exist: bright, dark, and planetary.

Bright (or diffuse) nebulae are frequently places of star formation. When stars begin to form, some are hot enough that their radiation excites the nebula's gas, ionizing it and causing it to shine. This is an emission nebula. If the stars aren't hot enough to cause ionization, the nebula's dust scatters and reflects their light, creating a reflection nebula.

The Orion Nebula (M42, Map 11), discovered in 1610, was the first bright nebula ever observed. And in 1780, M78 (Map 11), also in Orion, became the first reflection nebula discovered. Other emission nebulae in Messier's catalog (see p. 59 for more on this catalog) are the Lagoon Nebula (M8, Map 20), the Eagle

▲ **THE TRIFID NEBULA (M20)** combines emission and reflection nebulae, and includes lots of dust for good measure. ADAM BLOCK/MOUNT LEMMON SCIENCE CENTER/UNIVERSITY OF ARIZONA

◀ **THOR'S HELMET (NGC 2359)** is an emission nebula in Canis Major. The shape results from an intensely hot central star that's blowing the gas into space. ADAM BLOCK/MOUNT LEMMON SCIENCE CENTER/UNIVERSITY OF ARIZONA

45 Venus

If you observe Venus' phases during the daytime, use a yellow, orange, or red filter (darker ones on larger apertures) to enhance contrast and eliminate the blue component of the daytime sky.

Most planetary observers think a filter set helps to reveal details. Yellow, orange, and red filters also can cut the brightness of the daytime sky. MICHAEL E. BAKICH

44 Universal Time (UT)

Use it! Memorize your time zone's UT correction both with and without daylight saving time. Use UT in your observing log. Use it in your correspondence. Using UT is one way we all can standardize our observations.

46 Vitamin A

A diet deficient in vitamin A can lead to impaired night vision. An adequate intake of Vitamin A from foods such as eggs, cheese, liver, carrots, and most green vegetables will help ensure proper visual acuity at night. Please note, however, that *excessive* quantities of Vitamin A will not improve night vision and may be harmful to your health.

© WILDSTRAWBERRY_MAGIC | DREAMSTIME.COM

47 Weather and seeing

Some atmospheric factors indicate the quality or the steadiness of an astronomical image you will see. An air mass colder than the ground will produce puffy cumulus clouds and unsteady air, but it's usually relatively free of dust. An air mass warmer than the ground will produce stratiform clouds, haze, or mist and hold copious amounts of dust, but astronomical images will be steadier. Bad seeing is almost guaranteed at least 24 hours following the passage of a front (the boundary between warm and cool air masses) or trough (an elongated area of low pressure). Seeing can be very good with thin cirrus clouds aloft, but the opposite is true when high cirrus clouds combine with low-level crosswinds.

48 X-treme observing

Okay, it's a stretch, but few amateurs have X-ray telescopes. "X-treme" observing means viewing in cold weather. For low-temperature observing, preparation is everything. Pack more than you think you'll need. Ensuring you have a safe and comfortable observing session could hang on whether you brought along a single piece of gear. Most heat loss occurs from the head, so keep yours warm. My personal headgear consists of a fleece pullover head cover topped by a wide-brimmed hat. Heat also will seep into the cold ground through your boots. I decided to purchase the warmest boots that would allow me to easily

drive a vehicle. Hand warmers are superb but never seem to last the full time specified on the package. Keep warmers in your side pockets. Slip them in and out of your gloves or mittens for quick warm-ups. Finally, dress in layers. I generally wear fleece long underwear and thick pants. My upper body is covered with a T-shirt; a thin, long-sleeved, flannel shirt; a fleece pullover; and a down jacket. My wife, who is affected by the cold more than I am, wears a ski rescue suit as her outer layer. When fully zipped, with the hood up and boots and gloves in place, the wind has few places to chill her.

49 Your own pace

Some observers spend an hour or more on each object, trying to glean every bit of detail. Others take a leisurely pace of between five and fifteen objects per hour. Discover what works best for you.

50 Zoom eyepieces

If you don't have the cash to fill a large fishing tackle box with eyepieces, consider a high-quality zoom eyepiece. Such an accessory will provide you with a range of magnifications at a cost much less than the combined cost of each eyepiece within a single zoom lense's range of focal lengths.

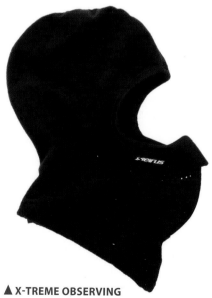

▲ **X-TREME OBSERVING**
A fleece pullover helps to retain heat where most of it is lost — your head.
MICHAEL E. BAKICH

INDEXES

The magnitude system

In the second century B.C., Greek astronomer Hipparchus of Rhodes cataloged a list of 850 stars into six brightness ranges, or magnitudes. The brightest stars he called 1st magnitude and the faintest 6th magnitude. Skywatchers used his system for 1,800 years.

Then, Italian astronomer Galileo Galilei noted his telescope revealed otherwise invisible objects. Galileo coined the term "7th-magnitude stars."

The invention of the telescope expanded the magnitude system. In the 18th century, a loose system defined two stars differing by 1 magnitude as having a brightness difference of 2.5 times.

In 1856, English astronomer Norman Pogson suggested a constant to calibrate all magnitudes. He changed the ratio between magnitudes from 2.5 to 2.5118865. This number, multiplied by itself five times, equals 100. So Sirius (Alpha Canis Majoris) at magnitude –1.47 is 100 times as bright as the star Wasat (Delta Geminorum) at magnitude 3.54, which is 5 magnitudes fainter.

Nineteenth-century astronomers noticed some stars are brighter than magnitude 1. The bright planets, the Moon, and the Sun are even brighter. So they defined magnitudes equal to and less than zero.

How bright a celestial object appears from Earth is its apparent magnitude (m). There's also a standardized — absolute — magnitude (M), which allows astronomers to compare the real brightnesses of objects. Absolute magnitude is the brightness a celestial object would have if its distance were 32.6 light-years (10 parsecs) from Earth.

Estimating limiting magnitude

While you're observing, it's a good idea to estimate limiting visual magnitude. This helps determine how good the conditions are and assigns a number value. Such an estimate also allows you to judge the quality of your observations months or years after that session. Most amateurs take a limiting-magnitude estimate near the zenith (the overhead point), where lights on the horizon are least obvious and there is less atmosphere to look through.

Another option is to make a limiting-magnitude estimate using the stars near Polaris (Alpha Ursae Minoris). Also, you can estimate telescopic limiting magnitude: Insert a medium-power (approximately 100x) eyepiece into your scope. Point your scope near, but not at, your target object. Sketch all visible stars in the area. Later, compare your drawing with a star chart or your computer monitor, if you use sky-plotting software.

THE GREEK ALPHABET AND STAR MAPS

Modern-day star designations (excluding proper names like Arcturus and Betelgeuse) date from 1603. In that year, German mapmaker Johannes Bayer published *Uranometria*, an atlas of the constellations. He plotted more than 2,000 stars, and his system differed from previous charts. Before Bayer, stellar cartographers designated stars by their positions within the mythological figures of the constellations.

Bayer's system used Greek letters to differentiate the brightnesses of stars in a constellation. So the first Greek letter, Alpha (α), usually denoted a constellation's brightest star, Beta (β) was the second-brightest, and so on. Bayer estimated brightnesses by eye, so some discrepancies exist.

Also, Bayer sometimes lettered stars sequentially. The Big Dipper is an example. Starting at the end of the bowl, Bayer lettered its stars Alpha, Beta, Gamma, Delta, Epsilon, Zeta, and Eta. Since Bayer's time, other celestial cartographers have subdivided some of his letters — for example, in Orion, we now find the stars π^1, π^2, π^3, etc. — but none have added new letters to any constellation.

ASTRONOMY EXTRA

Look on each star map for the following box:

α	Alpha	ι	Iota	ρ	Rho
β	Beta	κ	Kappa	σ	Sigma
γ	Gamma	λ	Lambda	τ	Tau
δ	Delta	μ	Mu	υ	Upsilon
ε	Epsilon	ν	Nu	φ	Phi
ζ	Zeta	ξ	Xi	χ	Chi
η	Eta	o	Omicron	ψ	Psi
θ	Theta	π	Pi	ω	Omega

CONSTELLATIONS

All stars are grouped into constellations — 88 of these patterns cover the sky. There's no overlap among constellations and no gaps between them. The International Astronomical Union made the constellation boundaries official in 1928. Today, when we talk about celestial objects being "in" a particular constellation, this means you'll find the object within that star group's official boundaries.

Constellation	Abbreviation	Pronunciation	Best seen	Map
Andromeda	And	an drah' meh dah	early Oct.	4
Antlia	Ant	ant' lee ah	late Feb.	18
Apus	Aps	ape' us	late May	23
Aquarius	Aqr	ah qwayr' ee us	late Aug.	15
Aquila	Aql	ak' wi lah	mid-July	14
Ara	Ara	air' ah	early June	24
Aries	Ari	air' eeze	late Oct.	10
Auriga	Aur	or eye' gah	late Dec.	5
Boötes	Boo	bow owe' teez	early May	7
Caelum	Cae	see' lum	early Dec.	17
Camelopardalis	Cam	kam uh low par' dah lis	late Dec.	1
Cancer	Can	kan' sir	late Jan.	6
Canes Venatici	CVn	kay' neez ven ah tee' see	early April	7
Canis Major	CMa	kay' nis may' jor	early Jan.	17
Canis Minor	CMi	kay' nis my' nor	mid-Jan.	11
Capricornus	Cap	kap ri kor' nus	early Aug.	21
Carina	Car	kah ree' nah	late Jan.	22
Cassiopeia	Cas	kass ee oh pee' uh	early Oct.	1
Centaurus	Cen	sen tor' us	late May	19
Cepheus	Cep	see' fee us	late Sept.	3
Cetus	Cet	see' tus	mid-Oct.	10
Chamaeleon	Cha	kah meel' ee an	early March	23
Circinus	Cir	sir sin' us	late April	23
Columba	Col	kol um' bah	mid-Dec.	17
Coma Berenices	Com	koe' mah bear uh nye' seez	early April	7
Corona Australis	CrA	kor oh' nah os tral' iss	late June	20
Corona Borealis	CrB	kor oh' nuh boar ee al' iss	mid-May	7
Corvus	Crv	kor' vus	late March	19
Crater	Crt	kray' ter	mid-March	12
Crux	Cru	kruks	late March	23
Cygnus	Cyg	sig' nus	late July	9
Delphinus	Del	del fee' nus	late July	15
Dorado	Dor	dor ah' doh	mid-Dec.	22
Draco	Dra	dray' koh	late May	2
Equuleus	Equ	ek woo oo' lee us	early Aug.	15
Eridanus	Eri	air uh day' nus	early Nov.	16
Fornax	For	for' nax	early Nov.	16
Gemini	Gem	gem' in eye	early Jan.	5
Grus	Gru	groose	late Aug.	21
Hercules	Her	her' cue leez	mid-Jan.	8
Horologium	Hor	hor uh low' gee um	early Nov.	22
Hydra	Hya	hi' drah	mid-March	12
Hydrus	Hyi	hi' drus	late Oct.	22
Indus	Ind	in' dus	mid-Aug.	24

Constellation	Abbreviation	Pronunciation	Best seen	Map
Lacerta	Lac	lah sir' tah	late Aug.	9
Leo	Leo	lee' owe	early March	12
Leo Minor	LMi	lee' owe my' nor	late Feb.	6
Lepus	Lep	lee' pus	mid-Dec.	17
Libra	Lib	lee' brah	early May	13
Lupus	Lup	loo' pus	early May	19
Lynx	Lyn	links	mid-Jan.	5
Lyra	Lyr	lye' rah	early July	8
Mensa	Men	men' sah	mid-Dec.	22
Microscopium	Mic	my kroh scop' ee um	early Aug.	21
Monoceros	Mon	mon oss' sir us	early Jan.	11
Musca	Mus	mus' kah	late March	23
Norma	Nor	nor' mah	mid-May	23
Octans	Oct	ok' tans	—	24
Ophiuchus	Oph	off ee oo' kus	mid-June	14
Orion	Ori	or eye' on	mid-Dec.	11
Pavo	Pav	pah' voh	mid-July	24
Pegasus	Peg	peg' ah sus	early Sept.	9
Perseus	Per	pur' see us	early Nov.	4
Phoenix	Phe	fee' niks	early Oct.	16
Pictor	Pic	pik' tor	mid-Dec.	22
Pisces	Psc	pye' seez	late Sept.	10
Piscis Austrinus	PsA	pye' sis os try' nus	late Aug.	21
Puppis	Pup	pup' iss	early Jan.	17
Pyxis	Pyx	pik' sis	early Feb.	18
Reticulum	Ret	reh tik' yoo lum	mid-Nov.	22
Sagitta	Sge	sah gee' tah	mid-July	8
Sagittarius	Sgr	sa ji tare' ee us	early July	20
Scorpius	Sco	skor' pee us	early June	20
Sculptor	Scl	skulp' tor	late Sept.	16
Scutum	Sct	skoo' tum	early July	14
Serpens	Ser	sir' pens	early June	14
Sextans	Sex	sex' tans	late Feb.	12
Taurus	Tau	tor' us	late Nov.	5
Telescopium	Tel	tel es koh' pee um	mid-July	20
Triangulum	Tri	try ang' yoo lum	late Oct.	4
Triangulum Australe	TrA	try ang' yoo lum os trail'	late May	23
Tucana	Tuc	too kan' ah	mid-Sept.	24
Ursa Major	UMa	ur' sah may' jor	mid-March	2
Ursa Minor	UMi	ur' sah my' nor	—	2
Vela	Vel	vay' lah	mid-Feb.	18
Virgo	Vir	vir' go	mid-April	13
Volans	Vol	voh' lans	mid-Jan.	22
Vulpecula	Vul	vul pek' yoo lah	late July	9

Note: The "Best seen" column gives the midnight culmination dates for the constellations' central points. These dates indicate when a constellation lies opposite the Sun and is highest at midnight. "Map" indicates where a constellation's central point (in R.A. and Dec.) lies. Parts of Octans and Ursa Minor surround the South and North celestial poles, respectively.

Beyond the constellation figures

In addition to the 88 constellations, figures called asterisms also exist. In the sky lore of some cultures, asterisms were coequal with constellations. Most asterisms are easier to recognize than many constellations because asterisms usually contain bright stars.

Today, we define an asterism as an unofficially recognized star pattern. This sets asterisms apart from constellations, which are star figures officially recognized by the International Astronomical Union.

We form some asterisms from single constellations. The Big Dipper — seven bright stars in Ursa Major — is the best example. But asterisms also may come from several constellations. The Winter Triangle is the Alpha stars of Orion, Canis Major, and Canis Minor.

Messier no.	NGC no.	Constellation	Type	Magnitude	Map
M1	1952	Taurus	SNR	8.0	5
M2	7089	Aquarius	GC	6.6	15
M3	5272	Canes Venatici	GC	6.3	7
M4	6121	Scorpius	GC	5.4	20
M5	5904	Serpens	GC	5.7	13
M6	6405	Scorpius	OC	4.2	20
M7	6475	Scorpius	OC	3.3	20
M8	6523	Sagittarius	N	6.0	20
M9	6333	Ophiuchus	GC	7.8	20
M10	6254	Ophiuchus	GC	6.6	14
M11	6705	Scutum	OC	5.8	14
M12	6218	Ophiuchus	GC	6.1	14
M13	6205	Hercules	GC	5.8	8
M14	6402	Ophiuchus	GC	7.6	14
M15	7078	Pegasus	GC	6.3	15
M16	6611	Serpens	N	6.0	14
M17	6618	Sagittarius	N	7.0	14
M18	6613	Sagittarius	OC	6.9	14
M19	6273	Ophiuchus	GC	6.8	20
M20	6514	Sagittarius	N	9.0	20
M21	6531	Sagittarius	OC	5.9	20
M22	6656	Sagittarius	GC	5.2	20
M23	6494	Sagittarius	OC	5.5	20
M24	6603	Sagittarius	SC	2.5	20
M25	IC 4725	Sagittarius	OC	4.6	20
M26	6694	Scutum	OC	8.0	14
M27	6853	Vulpecula	PN	7.3	8
M28	6626	Sagittarius	GC	6.9	20
M29	6913	Cygnus	OC	6.6	9
M30	7099	Capricornus	GC	6.9	21
M31	224	Andromeda	Gal	3.4	4
M32	221	Andromeda	Gal	8.2	4
M33	598	Triangulum	Gal	5.7	4
M34	1039	Perseus	OC	5.2	4
M35	2168	Gemini	OC	5.1	5
M36	1960	Auriga	OC	6.0	5
M37	2099	Auriga	OC	5.6	5
M38	1912	Auriga	OC	6.4	5
M39	7092	Cygnus	OC	4.6	9
M40	Win4	Ursa Major	Dbl	9.0/9.6	2
M41	2287	Canis Major	OC	4.5	17
M42	1976	Orion	N	3.7	11
M43	1982	Orion	N	6.8	11
M44	2632	Cancer	OC	3.1	6
M45	—	Taurus	OC	1.5	4
M46	2437	Puppis	OC	6.1	11
M47	2422	Puppis	OC	4.4	11
M48	2548	Hydra	OC	5.8	12
M49	4472	Virgo	Gal	8.4	13
M50	2323	Monoceros	OC	5.9	11
M51	5194	Canes Venatici	Gal	8.4	7
M52	7654	Cassiopeia	OC	6.9	3

Messier no.	NGC no.	Constellation	Type	Magnitude	Map
M53	5024	Coma Berenices	GC	7.7	7
M54	6715	Sagittarius	GC	7.2	20
M55	6809	Sagittarius	GC	6.3	20
M56	6779	Lyra	GC	8.4	8
M57	6720	Lyra	PN	8.8	8
M58	4579	Virgo	Gal	9.6	13
M59	4621	Virgo	Gal	9.6	13
M60	4649	Virgo	Gal	8.8	13
M61	4303	Virgo	Gal	9.6	13
M62	6266	Ophiuchus	GC	6.4	20
M63	5055	Canes Venatici	Gal	8.6	7
M64	4826	Coma Berenices	Gal	8.5	7
M65	3623	Leo	Gal	8.8	12
M66	3627	Leo	Gal	9.0	12
M67	2682	Cancer	OC	6.9	12
M68	4590	Hydra	GC	7.6	19
M69	6637	Sagittarius	GC	7.4	20
M70	6681	Sagittarius	GC	7.8	20
M71	6838	Sagitta	GC	8.0	8
M72	6981	Aquarius	GC	9.2	15
M73	6994	Aquarius	OC	8.9	15
M74	628	Pisces	Gal	8.5	10
M75	6864	Sagittarius	GC	8.6	21
M76	650	Perseus	PN	10.1	4
M77	1068	Cetus	Gal	8.9	10
M78	2068	Orion	N	8.0	11
M79	1904	Lepus	GC	7.7	17
M80	6093	Scorpius	GC	7.3	20
M81	3031	Ursa Major	Gal	6.9	2
M82	3034	Ursa Major	Gal	8.4	2
M83	5236	Hydra	Gal	7.5	19
M84	4374	Virgo	Gal	9.1	13
M85	4382	Coma Berenices	Gal	9.1	13
M86	4406	Virgo	Gal	8.9	13
M87	4486	Virgo	Gal	8.6	13
M88	4501	Coma Berenices	Gal	9.6	13
M89	4552	Virgo	Gal	9.7	13
M90	4569	Virgo	Gal	9.5	13
M91	4548	Coma Berenices	Gal	10.1	13
M92	6341	Hercules	GC	6.5	8
M93	2447	Puppis	OC	6.2	17
M94	4736	Canes Venatici	Gal	8.2	7
M95	3351	Leo	Gal	9.7	12
M96	3368	Leo	Gal	9.2	12
M97	3587	Ursa Major	PN	9.9	2
M98	4192	Coma Berenices	Gal	10.1	13
M99	4254	Coma Berenices	Gal	9.9	13
M100	4321	Coma Berenices	Gal	9.3	13
M101	5457	Ursa Major	Gal	7.9	2
M102	5866	Draco	Gal	10.0	2
M103	581	Cassiopeia	OC	7.4	1
M104	4594	Virgo	Gal	8.0	13
M105	3379	Leo	Gal	9.3	12
M106	4258	Canes Venatici	Gal	8.3	7
M107	6171	Ophiuchus	GC	7.8	14
M108	3556	Ursa Major	Gal	10.0	2
M109	3992	Ursa Major	Gal	9.8	2

KEY:

Dbl = Double star
Gal = Galaxy
GC = Globular cluster
N = Nebula

OC = Open cluster
PN = Planetary nebula
SC = Star cloud
SNR = Supernova remnant

The Messier objects — 109 deep-sky treats

Charles Messier (1730–1817) was a French comet-hunter. He occasionally encountered celestial objects that looked like comets because they appeared fuzzy in his small telescope. These objects didn't move against the starry background, however. Astronomers of the day called these objects nebulae, after the Latin word for "cloud," but the word has a more specific meaning today. The catalog Messier eventually compiled contains open and globular star clusters, galaxies, and bright and planetary nebulae.

Messier discovered his first non-moving objects in Taurus on August 28, 1758. This object became his first catalog entry: M1. Messier published three versions of his catalog. The first, in 1771, contained 45 objects. The second, with 68 objects, appeared in 1780. The third, containing 103 objects, appeared the next year. Later discoveries by Messier and others brought the final tally of objects to 109.

Messier marathons

Because the distribution of Messier's objects is uneven across the sky, a window exists when observers can see all 109 in a single night. Messier marathons have been popular with astronomy clubs since the early 1970s. One of the first extended mentions of the Messier marathon as an event appeared in the Forum section of the March 1980 issue of *Astronomy*, authored by California comet-hunter Don Machholz.

Tom Polakis of Arizona calculated the window of opportunity for a complete Messier marathon. The observing window begins on the date when globular cluster M30 is high enough to see in a dark morning sky. Working with the premise that the object has to be at an altitude of 2° for an observer to see it, Polakis calculated the Sun's altitude at that time and its distance from M30. On the evening end, the limiting object is spiral galaxy M74. The ability to spot it defines the end of the window of marathon observing dates.

These numbers are for 33° north latitude, and Polakis notes they get more favorable as you observe farther south, particularly for M30. The beginning of the season is March 17. Using the same criterion for the evening view of M74, the marathon season ends April 3.

Can you see them all?

Many observers have viewed all the Messier objects during a single night. You need a high-quality 3-inch or larger scope and a clear, dark, mid-northern-latitude site. Observe around New Moon. Some observers challenge themselves by trying to see as many objects as possible through small telescopes or binoculars.

You can find a list of the objects in marathon search order, which is not their standard numerical order. Don't rush, even through the close-knit Virgo galaxies. You'll have plenty of time to observe these objects. In fact, there's even a period — from about midnight to 3 A.M. local time — when you can get some sleep. Whether you take advantage of this or not depends on how much fun you're having.

The Caldwell objects

Amateur astronomers recognize M11, M20, and M31 as designations for deep-sky objects found on the list of 18th-century comet-hunter Messier. M11 is the Wild Duck Cluster in Scutum, M20 is the Trifid Nebula in Sagittarius, and M31 is the Andromeda Galaxy. But what if the letter C replaced the M in front of those numbers? Would you recognize C11, C20, and C31? This trio of deep-sky objects is also well known, but perhaps not by these designations.

The C stands for Caldwell or, more specifically, for Caldwell-Moore, the full surname of well-known British astronomy popularizer Sir Patrick Moore. When it came time to place an identifier by each of the numbered objects on his list, he couldn't use M for

Moore because Messier had taken that letter already, so he chose C. And in case you're still wondering, C11 is the Bubble Nebula (NGC 7635) in Cassiopeia, C20 is the North America Nebula (NGC 7000) in Cygnus, and C31 is the Flaming Star Nebula (IC 405) in Auriga.

Moore devised his list in 1995. Whereas Messier's list only includes objects visible from the latitude of Paris (48°51'), where Messier made his observations, Moore selected many southern objects. In fact, 34 Caldwell objects are invisible from Paris and five others never climb more than 3° above the southern horizon there.

Compared to Messier's list, Moore slightly reduced the number of certain types of objects while increasing others. He reduced the number of star clusters and galaxies slightly, but he increased the number of nebulae. Only four planetary nebulae made Messier's list, but Moore selected 13. Likewise, Moore increased the number of bright nebulae from Messier's five to 12.

Sagittarius and Virgo, which have 15 and 11 Messier objects, respectively, have only one Caldwell object each. Cassiopeia, Centaurus, and Cygnus each contain six Caldwell objects — the most in any constellation. Fifty of the 88 constellations contain Caldwell objects, as opposed to 34 that contain Messier ones.

/// CALDWELL OBJECTS — PART 1

Caldwell no.	NGC no.	Constellation	Type	Magnitude	Map
1	188	Cepheus	OC	8.1	3
2	40	Cepheus	PN	11.6	3
3	4236	Draco	Gal	9.7	2
4	7023	Cepheus	N	6.8	3
5	IC 342	Camelopardalis	Gal	9.2	1
6	6543	Draco	PN	8.8	3
7	2403	Camelopardalis	Gal	8.9	1
8	559	Cassiopeia	OC	9.5	1
9	Sh2–155	Cepheus	N	7.7	3
10	663	Cassiopeia	OC	7.1	1
11	7635	Cassiopeia	N	7.0	3
12	6946	Cepheus	Gal	9.7	3
13	457	Cassiopeia	OC	6.4	1
14	869/884	Perseus	OC	4.3	1
15	6826	Cygnus	PN	9.8	8
16	7243	Lacerta	OC	6.4	9
17	147	Cassiopeia	Gal	9.3	4
18	185	Cassiopeia	Gal	9.2	4
19	IC 5146	Cygnus	N	10.0	9
20	7000	Cygnus	N	6.0	9
21	4449	Canes Venatici	Gal	9.4	7
22	7662	Andromeda	PN	9.2	9
23	891	Andromeda	Gal	9.9	4
24	1275	Perseus	Gal	11.6	4
25	2419	Lynx	GC	10.4	5
26	4244	Canes Venatici	Gal	10.6	7
27	6888	Cygnus	N	7.5	9
28	752	Andromeda	OC	5.7	4
29	5005	Canes Venatici	Gal	9.8	7
30	7331	Pegasus	Gal	9.5	9
31	IC 405	Auriga	N	6.0	5
32	4631	Canes Venatici	Gal	9.3	7
33	6992/5	Cygnus	SNR	—	9
34	6960	Cygnus	SNR	—	9
35	4889	Coma Berenices	Gal	11.4	7
36	4559	Coma Berenices	Gal	9.8	7
37	6885	Vulpecula	OC	5.7	9
38	4565	Coma Berenices	Gal	9.6	7
39	2392	Gemini	PN	9.9	5
40	3626	Leo	Gal	10.9	6
41	Hyades	Taurus	OC	1.0	11
42	7006	Delphinus	GC	10.6	15
43	7814	Pegasus	Gal	10.5	10
44	7479	Pegasus	Gal	11.0	15
45	5248	Boötes	Gal	10.2	13
46	2261	Monoceros	N	10.0	11
47	6934	Delphinus	GC	8.9	15
48	2775	Cancer	Gal	10.3	12
49	2237–9	Monoceros	N	—	11
50	2244	Monoceros	OC	4.8	11
51	IC 1613	Cetus	Gal	9.0	10
52	4697	Virgo	Gal	9.3	13
53	3115	Sextans	Gal	9.1	12
54	2506	Monoceros	OC	7.6	12

KEY:	Gal = Galaxy	OC = Open cluster
	GC = Globular cluster	PN = Planetary nebula
	N = Nebula	SNR = Supernova remnant

Caldwell no.	NGC no.	Constellation	Type	Magnitude	Map
55	7009	Aquarius	PN	8.3	15
56	246	Cetus	PN	8.0	10
57	6822	Sagittarius	Gal	9.3	14
58	2360	Canis Major	OC	7.2	11
59	3242	Hydra	PN	8.6	18
60	4038	Corvus	Gal	11.3	19
61	4039	Corvus	Gal	13.0	19
62	247	Cetus	Gal	8.9	16
63	7293	Aquarius	PN	6.5	21
64	2362	Canis Major	OC	4.1	17
65	253	Sculptor	Gal	7.1	16
66	5694	Hydra	GC	10.2	19
67	1097	Fornax	Gal	9.2	16
68	6729	Corona Australis	N	9.7	20
69	6302	Scorpius	PN	12.8	20
70	300	Sculptor	Gal	8.1	16
71	2477	Puppis	OC	5.8	17
72	55	Sculptor	Gal	8.2	16
73	1851	Columba	GC	7.3	17
74	3132	Vela	PN	8.2	18
75	6124	Scorpius	OC	5.8	20
76	6231	Scorpius	OC	2.6	20
77	5128	Centaurus	Gal	7.0	19
78	6541	Corona Australis	GC	6.6	20
79	3201	Vela	GC	6.7	18
80	5139	Centaurus	GC	3.6	19
81	6352	Ara	GC	8.1	20
82	6193	Ara	OC	5.2	20
83	4945	Centaurus	Gal	9.5	19
84	5286	Centaurus	GC	7.6	19
85	IC 2391	Vela	OC	2.5	23
86	6397	Ara	GC	5.6	24
87	1261	Horologium	GC	8.4	22
88	5823	Circinus	OC	7.9	23
89	6087	Norma	OC	5.4	23
90	2867	Carina	PN	9.7	23
91	3532	Carina	OC	3.0	23
92	3372	Carina	N	6.2	23
93	6752	Pavo	GC	5.4	24
94	4755	Crux	OC	4.2	23
95	6025	Triangulum Australe	OC	5.1	23
96	2516	Carina	OC	3.8	22
97	3766	Centaurus	OC	5.3	23
98	4609	Crux	OC	6.9	23
99	Coal Sack	Crux	DN	—	23
100	IC 2944	Centaurus	N	4.5	23
101	6744	Pavo	Gal	9.0	24
102	IC 2602	Carina	OC	1.9	23
103	2070	Dorado	N	1.0	22
104	362	Tucana	GC	6.6	24
105	4833	Musca	GC	7.3	23
106	104	Tucana	GC	4.0	24
107	6101	Apus	GC	9.3	23
108	4372	Musca	GC	7.8	23
109	3195	Chamaeleon	PN	—	23

KEY:
DN = Dark nebula
Gal = Galaxy
GC = Globular cluster
N = Nebula
OC = Open cluster
PN = Planetary nebula
SNR = Supernova remnant

Moore numbered the objects in his catalog by declination. He started in the northernmost region of the sky with C1 (NGC 188, an open cluster in Cepheus) and ended near the South Celestial Pole with C109 (NGC 3195, a planetary nebula in Chamaeleon).

Moore's final stipulation was that each Caldwell object be observable through a 4-inch telescope from a dark location. In fact, you can observe many of the objects on Moore's list through binoculars and some you can see with your unaided eyes. Others are more difficult to see and although you can glimpse them through a 4-inch scope, you need a larger instrument to bring out their details.

	Name	Designation	Magnitude	Map
1	Sirius	Alpha Canis Majoris	−1.47	11
2	Canopus	Alpha Carinae	−0.72	17
3	Rigil Kentaurus	Alpha Centauri	−0.29	23
4	Arcturus	Alpha Boötis	−0.04	13
5	Vega	Alpha Lyrae	0.03	8
6	Capella	Alpha Aurigae	0.08	5
7	Rigel	Beta Orionis	0.12	11
8	Procyon	Alpha Canis Minoris	0.34	11
9	Achernar	Alpha Eridani	0.50	22
10	Betelgeuse	Alpha Orionis	0.58	11
11	Hadar	Beta Centauri	0.60	23
12	Altair	Alpha Aquilae	0.77	14
13	Aldebaran	Alpha Tauri	0.85	11
14	Acrux	Alpha Crucis	0.94	23
15	Spica	Alpha Virginis	1.04	13
16	Antares	Alpha Scorpii	1.09	20
17	Pollux	Beta Geminorum	1.15	5
18	Fomalhaut	Alpha Piscis Austrini	1.16	21
19	Deneb	Alpha Cygni	1.25	9
20	Mimosa	Beta Crucis	1.30	23
21	Regulus	Alpha Leonis	1.35	12
22	Adhara	Epsilon Canis Majoris	1.51	17
23	Castor	Alpha Geminorum	1.59	5
24	Shaula	Lambda Scorpii	1.62	20
25	Gacrux	Gamma Crucis	1.63	23
26	Bellatrix	Gamma Orionis	1.64	11
27	Elnath	Beta Tauri	1.65	5
28	Miaplacidus	Beta Carinae	1.68	23
29	Alnilam	Epsilon Orionis	1.70	11
30	Alnair	Alpha Gruis	1.74	21

► **MU CEPHEI** is a bright, red star that undergoes variations in brightness over time. GIUSEPPE DONATIELLO

▼ **THE WINTER CONSTELLATION** Orion the Hunter contains several stars that are among the sky's brightest.

MARC VAN NORDEN

THE REDDEST BRIGHT STARS

Designation	Right ascension	Declination	Magnitude	Color index	Map
R Scl	1h27m	–32°33'	5.8	3.9	16
TW Hor	3h12m	–57°19'	5.7	2.3	22
R Dor	4h37m	–62°05'	5.5	1.6	22
R Lep	4h59m	–14°48'	7.7	5.7	11
W Ori	5h05m	1°11'	6.2	3.5	11
CE Tau	5h32m	18°36'	4.4	2.1	5
W Pic	5h43m	–46°27'	7.8	4.8	17
Y Tau	5h46m	20°42'	7.0	3.0	5
BL Ori	6h25m	14°43'	6.2	2.4	11
UU Aur	6h37m	38°27'	5.3	2.6	5
NP Pup	6h54m	–42°22'	6.3	2.3	17
W CMa	7h08m	–11°55'	6.9	2.5	11
X Cnc	8h55m	17°14'	6.6	3.4	12
Y Hya	9h51m	–23°01'	6.6	3.8	18
X Vel	9h55m	–41°35'	7.2	4.3	18
AB Ant	10h12m	–35°19'	6.7	2.3	18
U Ant	10h35m	–39°34'	5.4	2.9	18
U Hya	10h38m	–13°23'	4.8	2.8	12
VY UMa	10h45m	67°25'	6.0	2.4	2
V Hya	10h52m	–21°15'	6.8	5.5	18
SS Vir	12h25m	0°48'	6.6	4.2	13
Y CVn	12h45m	45°26'	4.9	2.5	7
RY Dra	12h56m	65°59'	6.4	3.3	2
UY Cen	13h17m	–44°42'	6.9	2.8	19
V766 Cen	13h47m	–62°35'	6.5	2.0	23
X TrA	15h14m	–70°05'	5.8	3.6	23
TW Oph	17h30m	–19°28'	7.9	4.8	20
V Pav	17h43m	–57°43'	6.7	4.2	24
T Lyr	18h32m	36°59'	8.5	5.5	8
V450 Sct	18h33m	–14°52'	5.5	2.0	14
S Sct	18h50m	–7°54'	7.5	3.1	14
V Aql	19h04m	–5°41'	7.5	4.2	14
V1942 Sgr	19h19m	–15°55'	6.9	2.3	14
UX Dra	19h22m	76°34'	5.9	2.9	3
AQ Sgr	19h34m	–16°22'	7.3	3.4	14
RT Cap	20h17m	–21°19'	7.4	4.0	21
T Ind	21h20m	–45°01'	6.0	2.4	21
Y Pav	21h24m	–69°44'	6.4	2.8	24
V460 Cyg	21h42m	35°31'	6.1	2.5	9
Mu Cep	21h44m	58°47'	4.1	2.3	3
Pi¹ Gru	22h23m	–45°57'	6.6	2.0	21
RW Cep	22h23m	55°58'	6.7	2.3	3
TX Psc	23h46m	3°29'	5.0	2.6	15

The *New General Catalogue*

Printed in 1888, John L. E. Dreyer's *New General Catalogue* (NGC) is one of the prime references for bright deep-sky objects. The catalog contains 7,840 entries.

Some well-known NGC objects are the Andromeda Galaxy (NGC 224), the Orion Nebula (NGC 1976), and the Ring Nebula (NGC 6720). Most objects have more than a single designation. For example, these objects also are Charles Messier's catalog entries M31, M42, and M57, respectively.

You'll find all objects entered according to their right ascension. Lying at right ascension 23h57m — and thus numbered near the catalog's end — is one of the sky's most spectacular open clusters, NGC 7789 in Cassiopeia (Map 1). A 4-inch scope reveals 50 stars here, and a 12-inch reveals nearly 200 points of light.

STAR MAPS

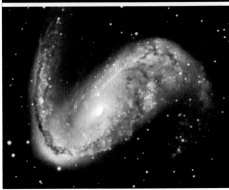

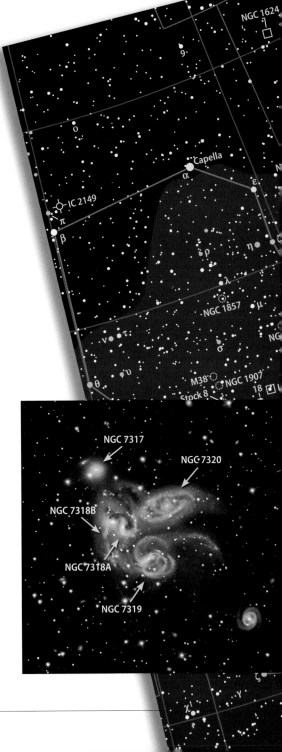

MAP 1 · NORTH POLAR 1

▲ **OPEN CLUSTER NGC 663** in Cassiopeia shines at magnitude 7.1 and measures 15' across. Use 50x or less to observe it.

The Queen's clusters

Cassiopeia the Queen is the highlight constellation of our first star map. Easily recognized by its W or M shape, many of the celestial targets in this area are open star clusters within Cassiopeia's boundaries. For more on how to observe these star groups, see "Observing open clusters" on page 46.

A notable sight outside Cassiopeia is **Kemble's Cascade** (NGC 1502), which glows at magnitude 5.7 in Camelopardalis the Giraffe. Franciscan amateur astronomer Father Lucian Kemble (1922–1999) first described this chance alignment of stars (it's not a true star cluster). He found it while scanning the sky through binoculars.

Binoculars that yield 15x reveal a dozen stars in a 2.5°-long chain. A small telescope shows 20 stars brighter than 11th magnitude, and through an 8-inch or larger scope, nearly 50 stars pop into view.

If you're new to observing, don't miss the **Double Cluster** (NGC 869 and NGC 884) in Perseus, which lies just to the southeast of Cassiopeia. You can see this bright pair of open clusters with naked eyes, but it's even better through binoculars. A telescope/eyepiece combination that magnifies 40x to 60x allows you to study pairs, triplets, and chains of stars within the clusters.

Two Messier objects inhabit Cassiopeia. **M103** lies 1° east of Delta Cassiopeiae, which is the left bottom star of the prominent W shape marking the Queen's throne. A 6-inch scope reveals three dozen stars grouped in a triangular area 5' across.

Draw a line from Alpha through Beta Cassiopeiae and extend it an equal distance to find **M52** (Map 3). Through an 8-inch scope, you'll see 75 stars clumped in various patterns. M52 lies on the edge of the Milky Way, so its stars aren't lost among the background points of light.

Also on Map 3, find the **Bubble Nebula** (NGC 7635). This object looks best through a 12-inch or larger telescope at a dark site. An 8-inch scope reveals a short arc of gas, but that's about it.

One more open cluster you shouldn't miss is the **Owl Cluster** (NGC 457), which lies less than 3° south of Delta Cassiopeiae. Even a 4-inch scope will show this cluster's two "wings." The eastern wing is a line of four bright stars, and the western wing comprises two pairs of stars arranged in a long rectangle.

You can find the bright spiral galaxy **NGC 2403** in Camelopardalis. Of course, in astronomy, "bright" is a relative term, but this galaxy makes the "top 100" brightest galaxies visible from Earth. Through a 6-inch scope, you'll see an oval patch 7' long with a central brightening.

▲ **SPIRAL GALAXY** NGC 2403 in Camelopardalis moves through space with M81, which lies in Ursa Major. NGC 2403 shines at magnitude 8.4 and lies 12 million light-years distant. FRED CALVERT/ADAM BLOCK/NOAO/AURA/NSF

▲ **THE DOUBLE CLUSTER IN PERSEUS** (NGC 869, right, and NGC 884) is visible to the naked eye, looks great through binoculars, and offers lots of enjoyment to telescopic observers. Use low power for a wide field of view.

▲ **THE OWL CLUSTER** (NGC 457) — also known as the ET Cluster — in Cassiopeia shines at magnitude 6.4 and is 20' across. Four stars on both the eastern and western edges form the owl's wings.

/// **DOUBLE-STAR DELIGHTS — MAP 1**

Designation	Right ascension	Declination	Magnitudes	Separation
STT 16	0h17m	54°39'	5.7, 10.2	13.3"
Upsilon Cassiopeiae	0h55m	58°58'	5.0, 12.5	14.3"
Psi Cassiopeiae	1h26m	68°07'	4.7, 8.9	25"
49 Cassiopeiae	2h06m	76°06'	5.3, 12.3	5.4"
Iota Cassiopeiae	2h29m	67°02'	4.6, 8.4	7.2"
1 Camelopardalis	4h32m	53°55'	5.7, 6.8	10.3"
3 Camelopardalis	4h40m	53°05'	5.3, 12.3	3.8"
5 Camelopardalis	4h55m	55°15'	5.6, 12.9	12.9"
29 Camelopardalis	5h51m	56°55'	6.5, 9.5	25.1"

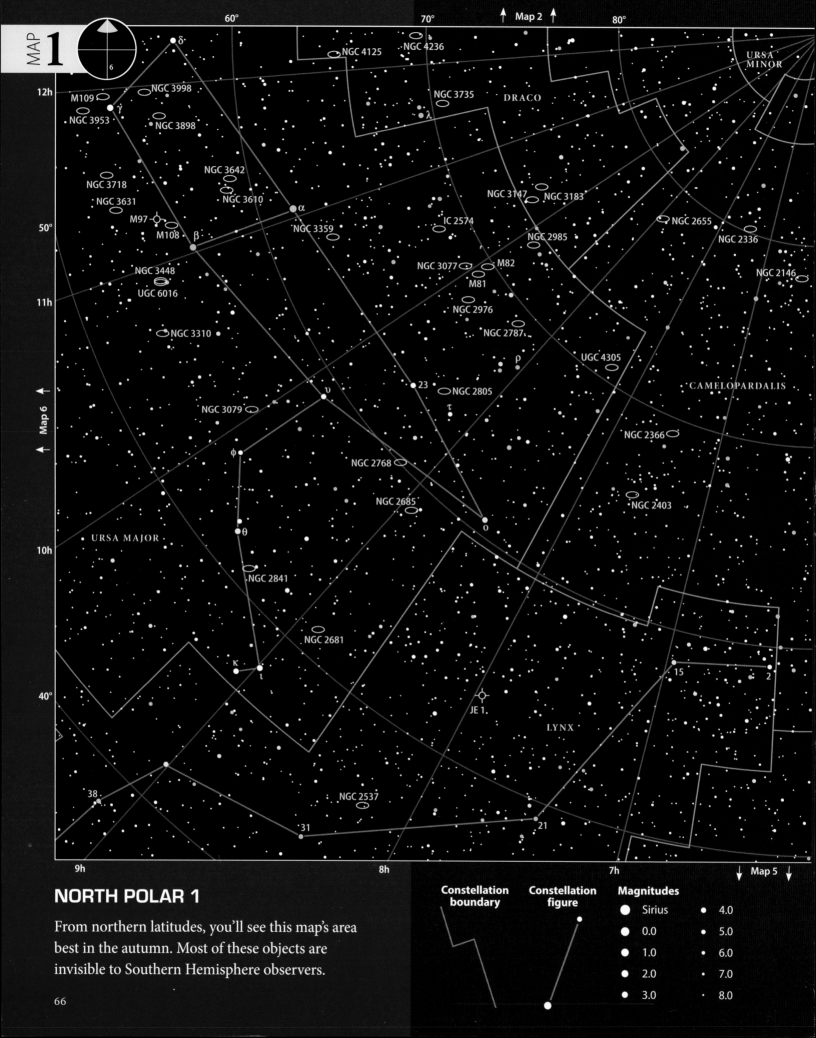

MAP 1

NORTH POLAR 1

From northern latitudes, you'll see this map's area
best in the autumn. Most of these objects are
invisible to Southern Hemisphere observers.

Constellation boundary

Constellation figure

Magnitudes

Sirius	4.0
0.0	5.0
1.0	6.0
2.0	7.0
3.0	8.0

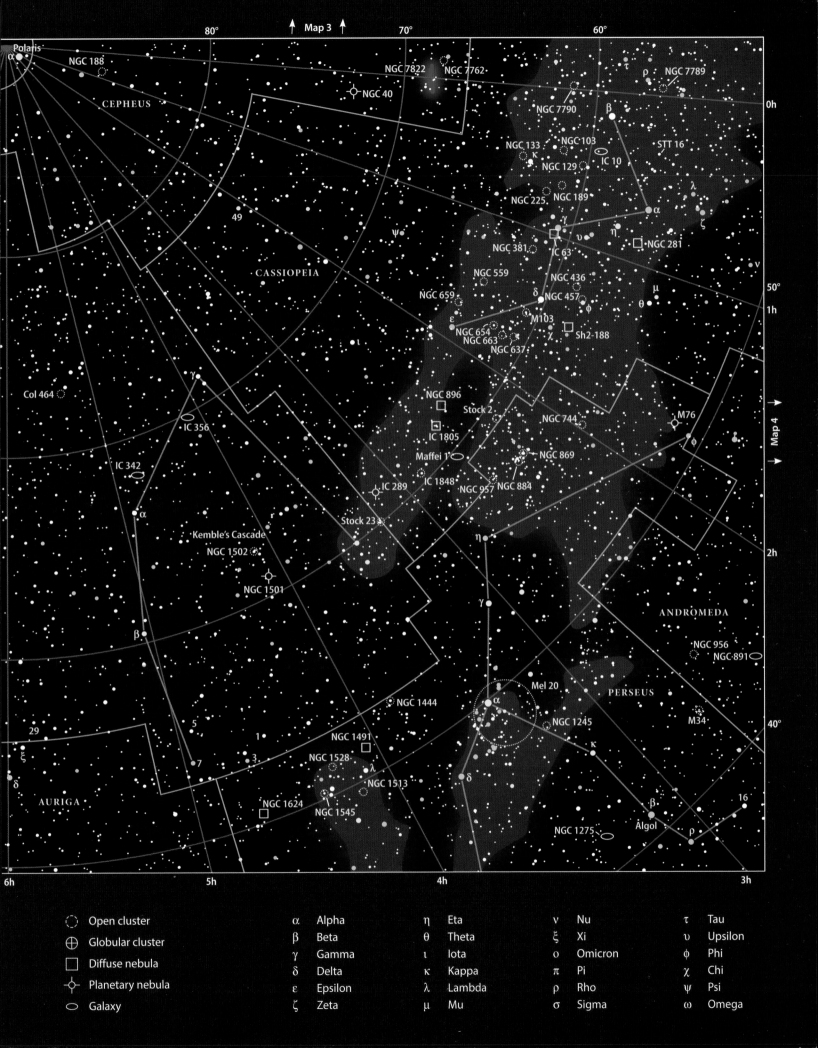

Polaris
α

NGC 188

CEPHEUS

NGC 7822
NGC 7762
NGC 40

NGC 7790
β

0h

τ   ρ   NGC 7789

NGC 133
κ   NGC 103
NGC 129   IC 10

STT 16

λ

49

ψ

NGC 225   NGC 189

γ   α   ζ

CASSIOPEIA

υ   η

NGC 381   IC 63   NGC 281

ν

NGC 559

50°

NGC 659   NGC 436
δ   NGC 457
M103   φ

μ
θ

ε   χ   Sh2-188

1h

ι   NGC 654
NGC 663   NGC 637

γ

Col 464

NGC 896

Stock 2

NGC 744   M76

IC 356

IC 1805

φ

IC 342

Maffei 1

NGC 869

α

IC 289   IC 1848

NGC 957   NGC 884

2h

Kemble's Cascade

Stock 23

η

NGC 1502

ANDROMEDA

NGC 1501

γ

NGC 956
NGC 891

β

NGC 1444

Mel 20
α

PERSEUS

5

1

NGC 1491

NGC 1245

40°

29

3

NGC 1528
λ

κ

ξ

7

NGC 1513

δ

16

δ

NGC 1624
NGC 1545

β
Algol

ρ

AURIGA

NGC 1275

M34

Map 4

⬭ Open cluster	α Alpha	η Eta	ν Nu	τ Tau
⊕ Globular cluster	β Beta	θ Theta	ξ Xi	υ Upsilon
☐ Diffuse nebula	γ Gamma	ι Iota	o Omicron	φ Phi
⬦ Planetary nebula	δ Delta	κ Kappa	π Pi	χ Chi
⬯ Galaxy	ε Epsilon	λ Lambda	ρ Rho	ψ Psi
	ζ Zeta	μ Mu	σ Sigma	ω Omega

MAP **2** | **NORTH POLAR 2**

▲ **BODE'S GALAXY (M81)**, left, and the Cigar Galaxy (M82) lie in the northernmost reaches of Ursa Major. Use an eyepiece that provides a wide field of view to see both of these galaxies at once. ANDREA TOSATTO

The Bear's realm

The dominant constellation on the next map is Ursa Major the Great Bear. Because this region is far from the Milky Way — where star clusters and nebulae abound — most of the great deep-sky objects here are galaxies. If you don't have a go-to telescope, the seven stars of the Big Dipper can help you find these objects.

Start with the galaxies on Messier's list: **Bode's Galaxy** (M81), the **Cigar Galaxy** (M82), **M101**, **M108**, and **M109**. An eyepiece/telescope combination that provides a field of view wider than ½° will catch both M81 and M82. Higher magnifications will reveal a large core and tight, graceful spiral arms in M81 and a wide, dark lane and splotchy bright areas in M82.

Many large spiral galaxies have up to a dozen nearby galaxies that, with the primary object, form a group. The M81 Group lies 11 million light-years away and counts about a dozen galaxies. To see them, use an 8-inch or larger scope from a dark site. On the other side of M81 from M82 is **NGC 3077**, a 10th-magnitude spiral.

It's worth finding the **Owl Nebula** (M97), a bright planetary nebula in Ursa Major. Under a dark sky, a 4-inch or larger telescope will show the Owl's face, which contains two round, dark areas (the "eyes"). The Owl is

▲ **ALTHOUGH M106 SHINES** at magnitude 8.4, it is an under-observed galaxy in Messier's catalog because it doesn't show much detail through small scopes. Look for its compact arms from a dark site. ADRIAN ZSILAVEC AND MICHELLE QUALLS/ADAM BLOCK/NOAO/AURA/NSF

a low-surface-brightness object that, except for a bit of a mottled surface, doesn't have a lot of detail.

The most famous double star in the sky lies at the bend of the Big Dipper's handle, and you don't need a telescope to see it. Alcor (80 Ursae Majoris) and Mizar (Zeta Ursae Majoris) once were a vision test for Roman soldiers. Mizar itself, however, is a much closer double star with a separation of only 14".

▲ **THE SPLINTER GALAXY** (NGC 5907) appears 10 times as long as wide because we see it edge-on. Also known as the Knife-edge Galaxy, NGC 5907 lies 40 million light-years away. BRAD EHRHORN/ADAM BLOCK/NOAO/AURA/NSF

▲ **LIKE A COSMIC** propeller, spiral galaxy M101 spins through space. You'll need a big scope to pick out the details of this low-surface-brightness galaxy. ADAM BLOCK/NOAO/AURA/NSF

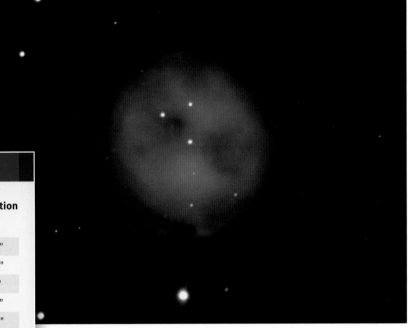

▲ **THE OWL NEBULA** (M97) resembles the bird after which it's named. This magnitude 9.9 planetary nebula measures 3.3' across. GARY WHITE AND VERLENNE MONROE/ADAM BLOCK/NOAO/AURA/NSF

/// DOUBLE-STAR DELIGHTS — MAP 2

Designation	Right ascension	Declination	Magnitudes	Separation
STT 188	8h22m	74°50'	6.5, 10.5	10.6"
23 Ursae Majoris	9h32m	63°03'	3.8, 9.0	22.8"
STF 1362	9h38m	73°05'	7.2, 7.2	4.7"
Upsilon Ursae Majoris	9h51m	59°03'	3.9, 11.5	11.3"
BU 1424	10h03m	50°07'	7.0, 7.2	28.2"
HJ 2534	10h33m	40°25'	4.8, 11.6	19.3"
Struve 1559	11h39m	64°22'	6.8, 7.8	2.0"
Zeta Ursae Majoris	13h24m	54°55'	2.3, 4.0	14.4"
STF 1774	13h40m	50°30'	6.4, 9.7	17.6"
DL Draconis	14h42m	61°15'	6.3, 8.5	4.1"
STF 1882	14h44m	61°05'	6.8, 8.3	11.6"
STF 1918	15h08m	63°07'	6.8, 10.8	17.8"

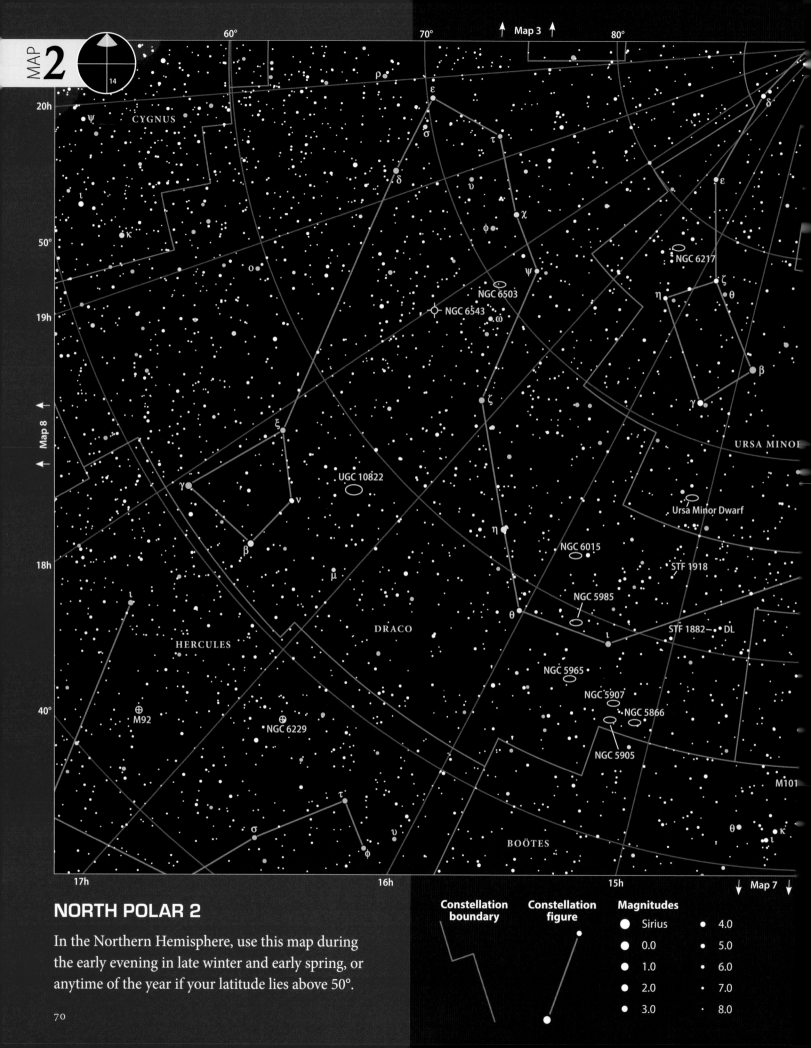

MAP
2

14

Map 3 ↑

20h
60°
70°
80°

CYGNUS
ρ
ε
ψ
σ
τ
ι
υ
χ
κ
φ
50°
NGC 6217
ψ
NGC 6503
ζ
θ
o
η
NGC 6543
ω
19h
ε
β
γ
Map 8
ξ
γ
URSA MINOR
UGC 10822
ν
Ursa Minor Dwarf
β
η
NGC 6015
STF 1918
18h
μ
NGC 5985
θ
STF 1882 — DL
ι
DRACO
NGC 5965
HERCULES
NGC 5907
M92
NGC 5866
40°
NGC 6229
NGC 5905
M101
τ
θ
κ
σ
υ
BOÖTES
φ
Map 7 ↓
17h
16h
15h

NORTH POLAR 2

In the Northern Hemisphere, use this map during
the early evening in late winter and early spring, or
anytime of the year if your latitude lies above 50°.

Constellation boundary	Constellation figure	Magnitudes	
		● Sirius	
		● 0.0	• 4.0
		● 1.0	• 5.0
		● 2.0	• 6.0
		● 3.0	• 7.0
			• 8.0

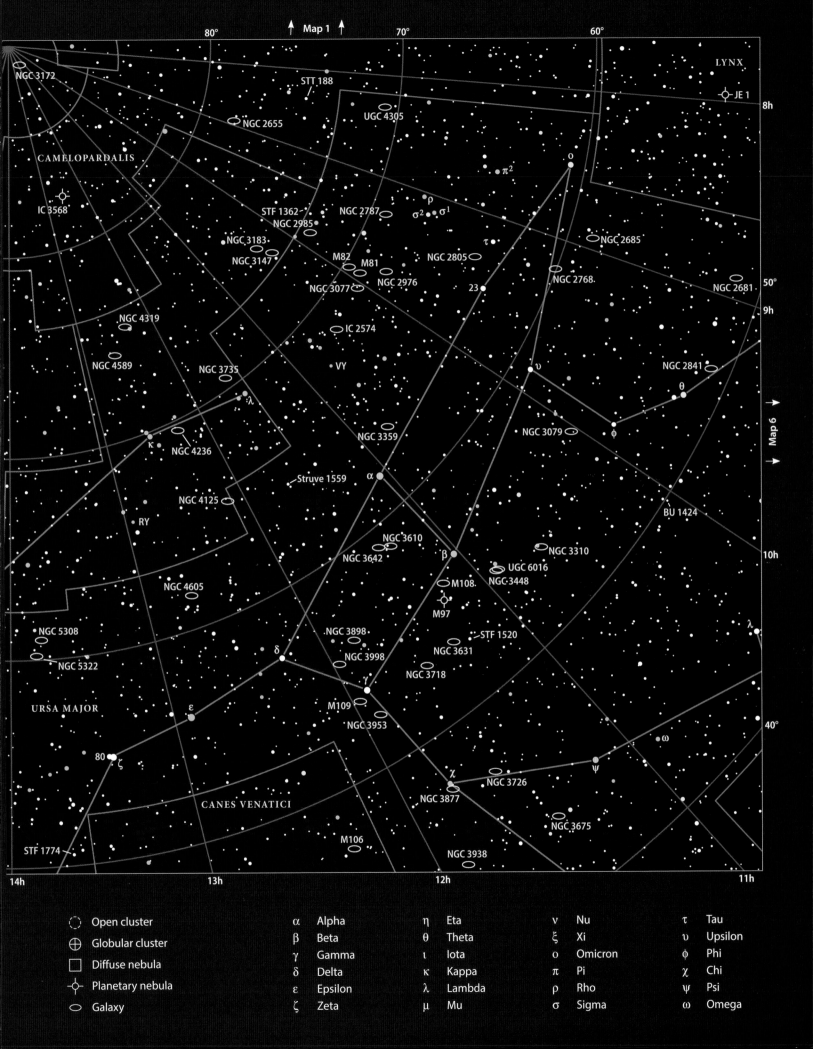

80°   70°   60°

LYNX

NGC 3172

JE 1   8h

CAMELOPARDALIS

STT 188

NGC 2655

UGC 4305

o

π²

IC 3568

ρ

NGC 2787

σ² σ¹

STF 1362

NGC 2685

NGC 2985

τ

NGC 3183

NGC 3147

M82   NGC 2805

NGC 2768   50°

M81

9h

NGC 3077   NGC 2976

NGC 4319

23   NGC 2681

IC 2574

NGC 4589

VY

NGC 3735

NGC 2841

υ

λ

θ

κ

NGC 3079

NGC 4236

φ

Struve 1559

α

BU 1424

NGC 4125

RY

10h

NGC 3610

NGC 4605

β

NGC 3310

NGC 3642

M108.   UGC 6016

NGC 3448

M97

NGC 5308

NGC 3898

STF 1520

λ

NGC 5322

NGC 3998   NGC 3631

δ

NGC 3718

γ

ω

ε

M109

40°

URSA MAJOR

NGC 3953

80

ζ

ψ

χ   NGC 3726

CANES VENATICI

NGC 3877

NGC 3675

STF 1774

M106

NGC 3938

14h   13h   12h   11h

⬡	Open cluster	α	Alpha	η	Eta
⊕	Globular cluster	β	Beta	θ	Theta
▢	Diffuse nebula	γ	Gamma	ι	Iota
⬦	Planetary nebula	δ	Delta	κ	Kappa
⬯	Galaxy	ε	Epsilon	λ	Lambda
		ζ	Zeta	μ	Mu

ν	Nu	τ	Tau
ξ	Xi	υ	Upsilon
o	Omicron	φ	Phi
π	Pi	χ	Chi
ρ	Rho	ψ	Psi
σ	Sigma	ω	Omega

Map 6 →→

MAP **3** NORTH POLAR 3

ADAM BLOCK/NOAO/AURA/NSF

▲ **NGC 6503 IN DRACO** is a lens-shaped spiral galaxy with compact arms. The star less than 4' east (left) of NGC 6503 is magnitude 8.6 SAO 8937.

Celestial sampler

Cepheus the King and part of the huge constellation Draco the Dragon make up most of Map 3. This area of sky may not contain many bright stars, but it offers a variety of celestial targets, from red stars to galaxies.

Start your deep-sky hunt 5° east of Zeta Draconis with the **Cat's Eye Nebula** (NGC 6543). Depending on your color perception, you'll either see this planetary nebula as greenish-blue or bluish-green, but you will see color. Through a 10-inch scope, the Cat's Eye resembles a somewhat unstructured spiral galaxy. An outer shell of gas 5' across surrounds the "eye," but seeing this feature requires a 16-inch or larger telescope.

Only 1.8° northeast of Omega Draconis lies **NGC 6503**, a magnitude 10.1 spiral galaxy that even a 4-inch scope will reveal. NGC 6503 is roughly three times as long as it is wide. Its central region appears elongated, but the brightest part does not sit over NGC 6503's center.

In Cepheus, you'll find open cluster **NGC 6939** easily. It forms an equilateral triangle with 3rd-magnitude Eta and 4th-magnitude Theta Cephei. NGC 6939 is a magnitude 7.8 group of roughly 60 stars between magnitudes 11 and 13 in an area 5' across.

Along Cepheus' southern edge lies **Herschel's Garnet Star** (Mu Cephei). Astronomer William Herschel first described it in the 18th century. Glowing at 4th magnitude, Mu is one of the sky's reddest stars — a sure cure for the observing blues.

Mu marks the northern edge of emission nebula **IC 1396**. This giant object spans 2° (four Full Moons) and is difficult to see through 6-inch or smaller telescopes. Through larger instruments, IC 1396 appears as a circular mist crossed by many dark lanes.

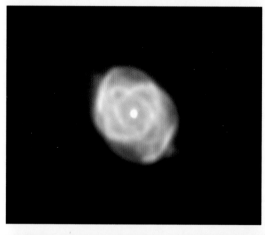

◄ **THE EYE IN THE SKY** in this case is the Cat's Eye Nebula (NGC 6543) in Draco. You can observe this magnitude 8.1 planetary nebula through any size telescope. NGC 6543 measures 20" **across.** ADAM BLOCK/ NOAO/AURA/NSF

ADAM BLOCK/NOAO/AURA/NSF

▲ **THE ELEPHANT TRUNK NEBULA** (van den Bergh 142) is a dark nebula that is part of the much larger IC 1396 nebula complex.

NGC 7538 lies 43' west-northwest of the larger and fainter Bubble Nebula (NGC 7635). An Oxygen-III filter will help you observe this object.

▶ THE BUBBLE NEBULA (NGC 7635) floats about 0.6° south-west of the bright star cluster M52. NGC 7635 glows because of an energetic star at its center. BRAD EHRHORN/ADAM BLOCK/NOAO/AURA/NSF

FRED CALVERT/ADAM BLOCK/NOAO/AURA/NSF

▲ OPEN CLUSTER NGC 7380 makes up one part of this fascinating sky region. The remainder is the bright emission nebula Sharpless 2–142.

KRIS SANDBURG AND PETER JACOBS/ADAM BLOCK/NOAO/AURA/NSF

/// DOUBLE-STAR DELIGHTS — MAP 3

Designation	Right ascension	Declination	Magnitudes	Separation
Eta Draconis	16h24m	61°30'	2.9, 8.9	5.2"
17 Draconis	16h36m	52°55'	5.6, 6.6	3.2"
41 Draconis	18h00m	79°59'	5.8, 6.2	19.1"
39 Draconis	18h24m	58°48'	4.9, 7.9	3.8"
Epsilon Draconis	19h48m	70°16'	4.0, 7.6	3.1"
65 Draconis	20h02m	64°37'	6.2, 10.1	5.6"
Kappa Cephei	20h09m	77°43'	4.4, 8.4	7.3"
Beta Cephei	21h29m	70°03'	3.3, 8.0	13.4"
Xi Cephei	22h04m	64°37'	4.4, 6.5	7.7"
19 Cephei	22h05m	62°17'	5.2, 11.2	19.8"
2 Cassiopeiae	23h10m	59°19'	5.6, 13.0	19.9"
Omicron Cephei	23h19m	68°06'	5.0, 7.6	3.2"
Sigma Cassiopeiae	23h59m	55°45'	5.0, 7.1	3.1"

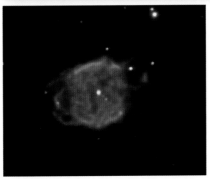

◀ THE BOW-TIE NEBULA (NGC 40) doesn't give up its secrets easily. Use an 8-inch or larger telescope from a dark site, and you'll note NGC 40's reddish hue.

STEVE AND PAUL MANDEL/ADAM BLOCK/NOAO/AURA/NSF

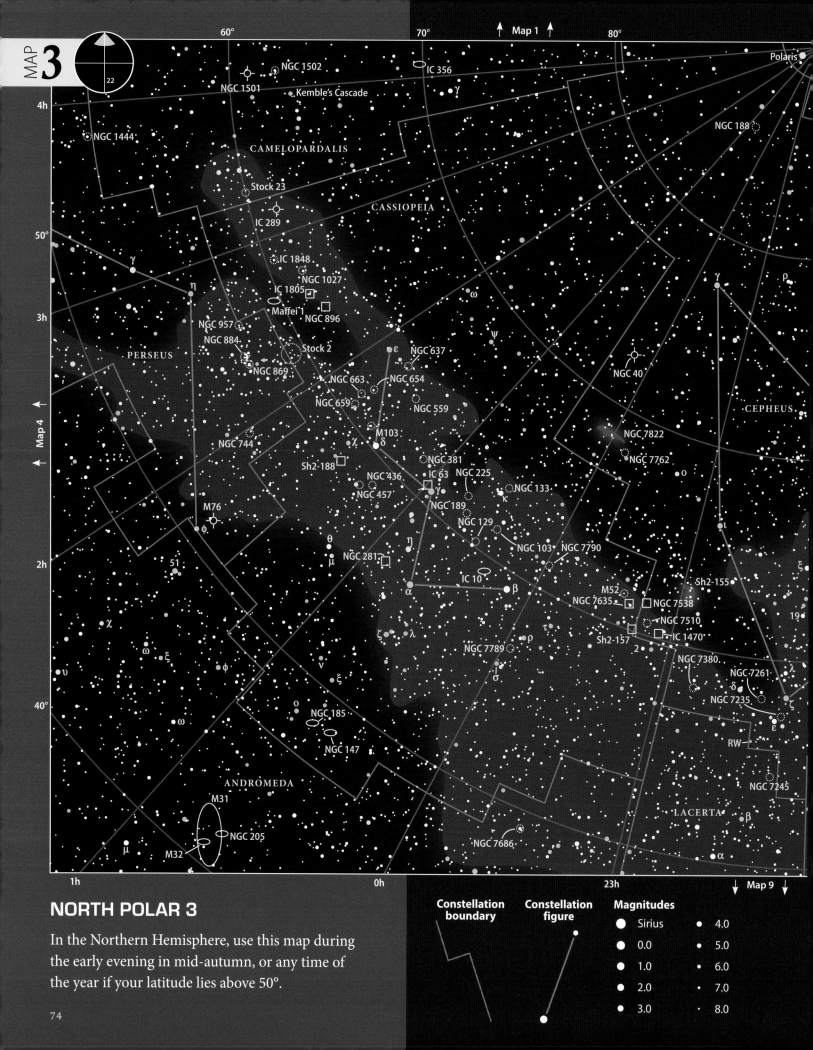

MAP **3**
22

↑ Map 1 ↑

60°   70°   80°

Polaris

4h

NGC 1502
NGC 1501
Kemble's Cascade
IC 356
γ
NGC 188

NGC 1444
CAMELOPARDALIS
NGC 1027

Stock 23
CASSIOPEIA
50°
IC 289
γ
ω
γ
ρ

γ
IC 1848
NGC 1027
NGC 7822
CEPHEUS

3h
η
IC 1805
Maffei 1
NGC 896
ψ
NGC 40
NGC 7762

NGC 957
NGC 884
Stock 2
ε
NGC 637
NGC 663   NGC 654
o

PERSEUS
NGC 869
NGC 659
NGC 559

NGC 744
M103
χ   δ
NGC 381
NGC 7822
ι

Sh2-188
IC 63   NGC 225
NGC 7762

Map 4
NGC 436
NGC 133
ξ
NGC 457
κ
19

M76
θ
NGC 189
NGC 129

φ
μ
η
NGC 103   NGC 7790
Sh2-155

2h
NGC 281
α
IC 10   β
M52
NGC 7635
NGC 7538

51
ζ   λ
τ
NGC 7510
IC 1470

χ
ρ
Sh2-157   2
λ

ω
ξ
ν
ξ
NGC 7789
σ
NGC 7380
NGC 7261
δ
NGC 7235

υ
ω
o
NGC 185
ε

40°
ω
NGC 147
RW

ANDROMEDA
NGC 7245

M31
LACERTA
β

μ
NGC 205
NGC 7686
α

M32

1h   0h   23h   ↓ Map 9 ↓

NORTH POLAR 3

In the Northern Hemisphere, use this map during
the early evening in mid-autumn, or any time of
the year if your latitude lies above 50°.

Constellation boundary	Constellation figure	Magnitudes	
		Sirius	4.0
		0.0	5.0
		1.0	6.0
		2.0	7.0
		3.0	8.0

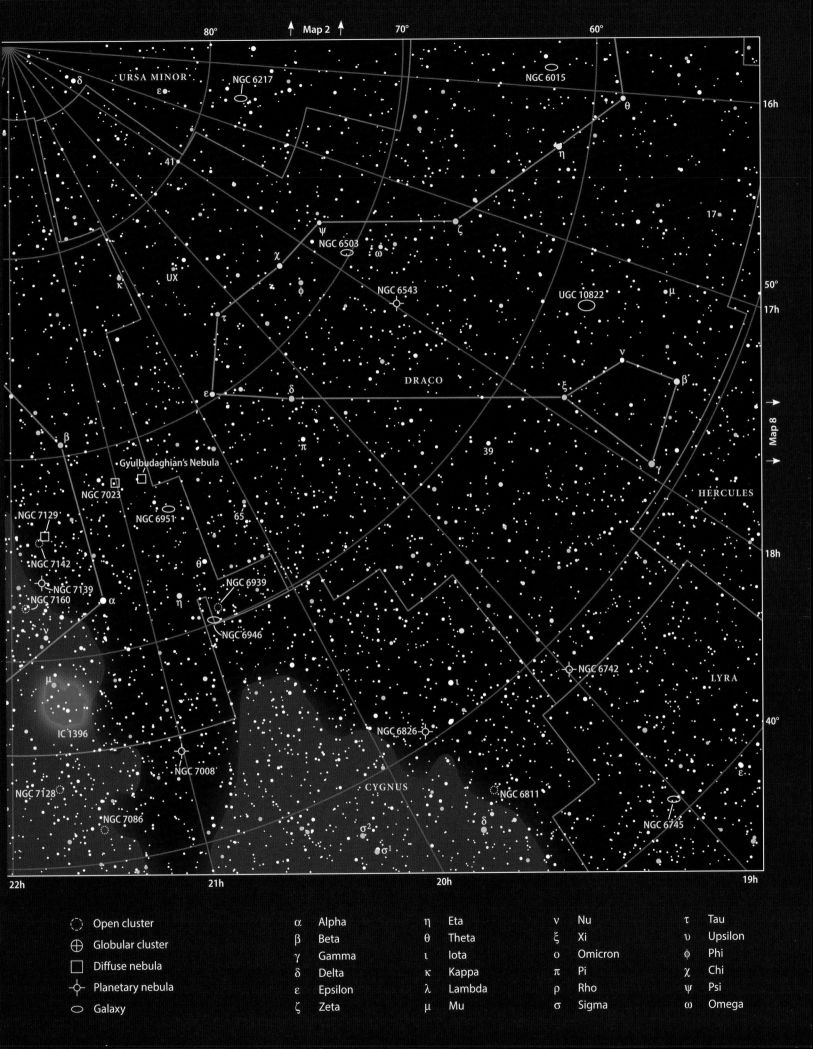

80°   70°   60°

16h

URSA MINOR

NGC 6217

NGC 6015

δ
ε

θ

41

η

17

ψ
NGC 6503

ζ

50°

χ

ω

17h

UX

z

φ

NGC 6543

UGC 10822

μ

κ

τ

DRACO

ν

β

ε   δ

ξ

γ

Map 8

β

π

39

HERCULES

Gyulbudaghian's Nebula

NGC 7023

NGC 6951

65

NGC 7129

θ

18h

NGC 7142

η

NGC 6939

NGC 7139

α

NGC 6746

NGC 7160

NGC 6946

ν

NGC 6742

LYRA

μ

ε

IC 1396

NGC 6826

40°

NGC 7008

CYGNUS

NGC 6811

NGC 7128

NGC 6745

NGC 7086

σ²

δ

σ¹

22h   21h   20h   19h

⬭ (dashed circle)	Open cluster
⊕	Globular cluster
□	Diffuse nebula
⊙	Planetary nebula
⬭	Galaxy

α	Alpha	η	Eta	ν	Nu	τ	Tau
β	Beta	θ	Theta	ξ	Xi	υ	Upsilon
γ	Gamma	ι	Iota	ο	Omicron	φ	Phi
δ	Delta	κ	Kappa	π	Pi	χ	Chi
ε	Epsilon	λ	Lambda	ρ	Rho	ψ	Psi
ζ	Zeta	μ	Mu	σ	Sigma	ω	Omega

MAP 4   NORTH EQUATORIAL 1

▲ SPIRAL GALAXY NGC 891 in Andromeda appears nearly edge-on from our vantage point. Through an 8-inch or larger telescope, you'll see the thin dust lane that bisects NGC 891 lengthwise. UCL/ULO/IAN HOWARTH (UCL PHYSICS & ASTRONOMY)

/// DOUBLE-STAR DELIGHTS — MAP 4

Designation	Right ascension	Declination	Magnitudes	Separation
Delta Andromedae	0h39m	30°52'	3.5, 13.0	28.7"
39 Andromedae	1h03m	41°20'	5.9, 12.3	20.1"
Gamma Arietis	1h54m	19°17'	4.8, 4.8	7.8"
Epsilon Trianguli	2h03m	33°16'	5.4, 11.4	4.0"
Gamma Andromedae	2h04m	42°20'	2.3, 5.5	9.8"
59 Andromedae	2h11m	39°02'	6.0, 6.7	16.7"
6 Trianguli	2h12m	30°18'	5.2, 6.6	3.9"
33 Arietis	2h41m	27°03'	5.5, 8.4	28.6"
Theta Persei	2h44m	49°13'	4.2, 10.0	19.7"
Pi Arietis	2h49m	17°28'	5.3, 8.8	3.2"
41 Arietis	2h50m	27°15'	3.7, 10.8	24.6"
20 Persei	2h54m	38°20'	5.3, 10.0	15.0"
STF 401	3h31m	27°33'	6.6, 6.9	11.1"
Sigma Persei	3h31m	47°51'	4.4, 6.2	22.9"
STF 431	3h42m	33°57'	5.0, 10.0	20.0"
Zeta Persei	3h54m	31°52'	2.9, 9.5	12.8"
Struve 469	3h57m	41°52'	6.8, 10.3	8.9"
Epsilon Persei	3h58m	40°00'	3.0, 8.2	8.8"
Mu Persei	4h15m	48°24'	4.3, 11.8	14.8"

The Princess' sky

The next map features two large constellations, Andromeda the Princess and Perseus the Hero, and two small ones, Triangulum the Triangle and Aries the Ram. All four have bright sections that make them easy to find.

Andromeda contains the ultimate northern deep-sky object: the **Andromeda Galaxy** (M31). Find it 1° west of Nu Andromedae. Visible to the naked eye even from moderately bright locations, M31 measures more than 3° long — or six Full Moons side by side.

The Andromeda Galaxy is a showpiece through any telescope. Small scopes with eyepieces that give a wide field of view let you study the galaxy's overall structure. Move up to a 6-inch scope at a dark site, and you'll see two dust lanes. Larger telescopes allow you to crank up the magnification and study individual features.

M31's two bright companions, **M32** and **NGC 205**, are small elliptical galaxies. Such objects normally appear featureless, and you'll notice this a lot more because of their positions near M31.

If you can tear your gaze from the Andromeda Galaxy, your reward will be a colorful sight in Andromeda: the **Blue Snowball** (NGC 7662), a planetary nebula lying 3,000 light-years away from Earth. Low-power views bring out NGC 7662's color best. More magnification reveals rich structure. A hollow region surrounds the 13th-magnitude central star.

Take a good look at Gamma Andromedae, a blue-and-orange double star, then move 3½° east to the spiral galaxy **NGC 891**. Veering only 1.4° from being exactly edge-on, NGC 891 is four times as long as it is wide, has a large, bright nucleus, and splits lengthwise because of a dark dust lane. Point a 10-inch or larger scope at NGC 891, and you'll see why it makes so many "top 10" lists of best-to-observe galaxies.

If you're observing from a light-polluted site, point your telescope 4.3° west of magnitude 3.4 Alpha Trianguli to find the **Pinwheel Galaxy** (M33). From a dark site, no such directions are required because M33 is visible to the naked eye.

Through a telescope, M33 explodes into detail, with multiple spiral arms, bright stellar associations, and — through a 12-inch or larger telescope — the emission nebula NGC 604. This object sits at the tip of M33's northern spiral arm. A nebula filter will highlight NGC 604.

At the map's eastern edge lies the **Pleiades** (M45), a naked-eye star cluster and, at magnitude 1.5, the brightest object in Messier's catalog. As you scan M45, let your telescopic gaze fall on Merope (23 Tauri). Surrounding this star is the Merope Nebula (NGC 1435), a cloud of gas passing through M45 and lit by its bright stars. NGC 1435 doesn't respond to nebula filters because it's a reflection nebula.

▲ THE PINWHEEL GALAXY (M33) lies in Triangulum and is more difficult to observe than its magnitude 5.7 brightness indicates. Because it faces us, M33's light spreads out over an area 73' by 45'. JARED SMITH

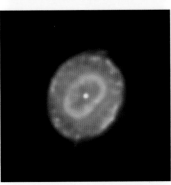

▲ THE ANDROMEDA GALAXY (M31) is one of the showpiece deep-sky objects. Whatever size telescope (or binoculars) you use, this object will reward you for the time you spend observing it. STEPHEN RAHN

◀ THE BLUE SNOWBALL (NGC 7662) is a planetary nebula in Andromeda. Its name is well-earned. Use 50x to 100x to bring out the color. ADAM BLOCK/NOAO/AURA/NSF

▲ THE CALIFORNIA NEBULA (NGC 1499) in Perseus looks much better in photographs than it does to the eye because of its low surface brightness. A Hydrogen-beta filter increases the contrast. STEPHEN RAHN

◀ THE PLEIADES (M45) is one of the closest star clusters. Some observers test their eyesight by counting Pleiads. How many can you see? ANDREA TOSATTO

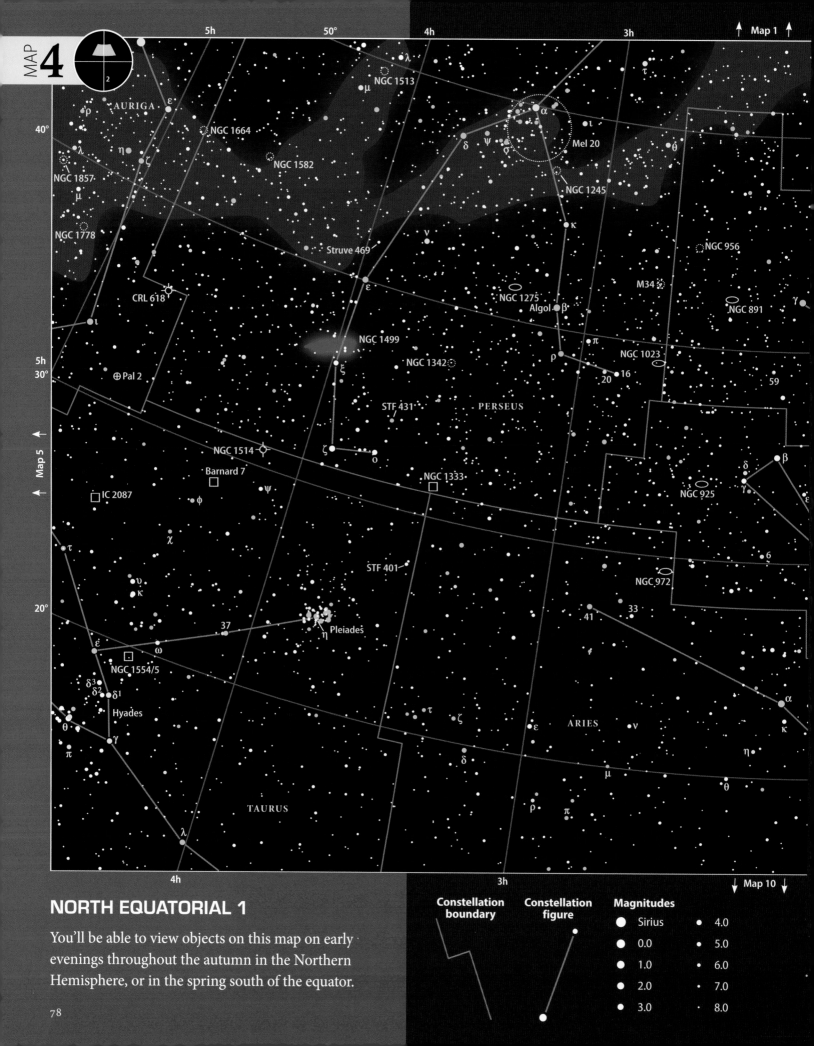

MAP
4

↑ Map 1 ↑

5h 50° 4h 3h

NGC 1513

AURIGA ε λ

ρ μ

40° NGC 1664 α ι

λ η ζ δ ψ Mel 20

NGC 1857 NGC 1582 σ θ

μ NGC 1245

NGC 1778 κ

NGC 956

ν M34

Struve 469 NGC 891

ε NGC 1275

CRL 618 Algol β

ι π

NGC 1499 ρ NGC 1023

5h ξ 16

30° NGC 1342 20 59

⊕ Pal 2

STF 431 PERSEUS

β

NGC 1514 ζ ο δ

Map 5 Barnard 7 NGC 1333 γ

φ NGC 925

IC 2087 ψ

ε

τ

χ 6

STF 401

υ NGC 972

κ 33

20° 37 41

η Pleiades

ω

ε

NGC 1554/5 τ α

δ³ ζ κ

δ² δ¹ ε ARIES ν

Hyades η

θ δ

π μ θ

γ ρ π

λ

TAURUS

4h 3h

↓ Map 10 ↓

NORTH EQUATORIAL 1

You'll be able to view objects on this map on early
evenings throughout the autumn in the Northern
Hemisphere, or in the spring south of the equator.

Constellation boundary	Constellation figure	Magnitudes	
		● Sirius	· 4.0
		● 0.0	· 5.0
		● 1.0	· 6.0
		● 2.0	· 7.0
		● 3.0	· 8.0

MAP **5** **NORTH EQUATORIAL 2**

MARC VAN NORDEN

▲ **THE CRAB NEBULA (M1)** is an expanding supernova remnant that appeared as a brilliant new "star" in Taurus in A.D. 1054.

All around Auriga

Late fall and early winter evenings in the Northern Hemisphere offer three constellations with a wide range of deep-sky objects: Auriga, Taurus, and Gemini. The Milky Way's rich star clouds pass right through Auriga and Gemini and skirt Taurus' eastern edge.

Charles Messier cataloged three objects in Auriga — all open clusters — that fall into a line that runs parallel to the Milky Way. The western-most is **M38**, which measures 20' across and shines at magnitude 6.4. Through an 8-inch scope, you'll be able to see more than 100 stars. Another open cluster, magnitude 8.2 **NGC 1907**, lies ½° south of M38 and forms a pair with it that's somewhat reminiscent of the Double Cluster in Perseus (see Map 1 or Map 3).

M36, at magnitude 6.0, and **M37**, which shines at magnitude 5.6, complete the triumvirate. You can see all three clusters with your naked eyes from a dark site. Through a telescope, M38 and M36 are excellent targets, but M37 truly is spectacular. A small scope reveals 50 stars in an area 10' across. Through a 12-inch scope, you'll see several hundred stars

filling the field of view of a medium-power eyepiece.

Many deep-sky objects in Taurus require at least an 8-inch telescope for you to appreciate them. The exceptions are Taurus' two Messier targets, the **Pleiades** (M45) and the **Crab Nebula** (M1). The Crab Nebula is the best-known supernova remnant in the sky. Find it 1° northwest of 3rd-magnitude Zeta Tauri. Measuring 6' by 4', M1 looks like a notched puff of smoke through small telescopes.

Gemini the Twins contains only one Messier object, magnitude 5.1 **M35**. This open cluster's central 20' contains more than 150 stars, and nearby there's a surprise: a smaller, more condensed star cluster, **NGC 2158**. You'll need to use a high-magnification eyepiece in a large scope to break this magnitude 8.6 object into individual stars.

An easy-to-spot planetary nebula in this region is the **Eskimo Nebula** (NGC 2392), also called the Clown-Face Nebula. Shining at magnitude 9.1, this planetary has a double-shell appearance visible through 8-inch and larger telescopes. NGC 2392 measures roughly 1' across.

Designation	Right ascension	Declination	Magnitudes	Separation
Struve 479	4h01m	23°11'	6.6, 9.0	8.0"
STT 72	4h08m	17°20'	7.4, 8.0	4.7"
Chi Tauri	4h23m	25°38'	5.5, 7.6	19.5"
Omega Aurigae	4h59m	37°53'	5.1, 8.1	5.3"
5 Aurigae	5h00m	39°24'	6.0, 9.7	3.8"
9 Aurigae	5h07m	51°36'	5.0, 12.2	5.2"
14 Aurigae	5h15m	32°40'	5.2, 7.4	14.6"
16 Aurigae	5h18m	33°22'	4.8, 10.6	4.2"
Struve 680	5h19m	20°07'	6.1, 9.6	9.0"
18 Aurigae	5h19m	33°59'	6.5, 11.8	4.1"
Sigma Aurigae	5h25m	37°23'	5.2, 11.2	8.7"
26 Aurigae	5h39m	30°30'	5.5, 8.0	12.4"
Theta Aurigae	5h59m	37°12'	2.7, 7.2	3.5"
41 Aurigae	6h12m	48°42'	6.1, 6.8	7.7"
59 Aurigae	6h53m	38°51'	6.2, 9.5	22.3"
STT 165	7h08m	15°56'	5.6, 11.3	10.4"
HO 343	7h14m	24°53'	6.0, 12.7	23.9"
Lambda Geminorum	7h18m	16°32'	3.6, 10.7	9.6"
Delta Geminorum	7h20m	21°59'	3.5, 8.5	6.2"
BU 1413	7h22m	20°26'	5.2, 12.2	16.5"
65 Aurigae	7h22m	36°45'	5.2, 11.7	11.4"
Rho Geminorum	7h29m	31°46'	4.2, 12.5	3.4"
Kappa Geminorum	7h44m	24°23'	3.7, 8.2	7.1"
Pi Geminorum	7h48m	33°25'	5.3, 11.4	21.0"

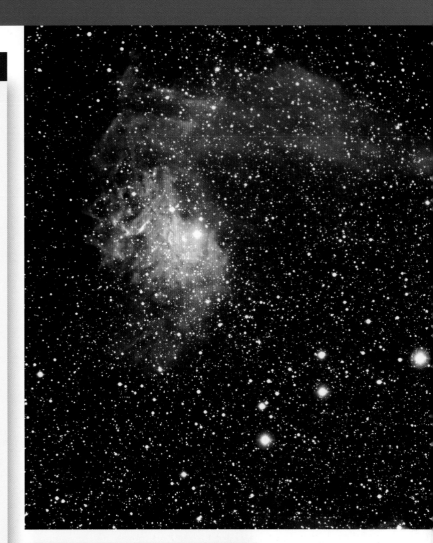

▲ THE FLAMING STAR NEBULA (IC 405) in Auriga (left of center) glows because of intense radiation from the star AE Aurigae.

ADAM BLOCK/NOAO/AURA/NSF

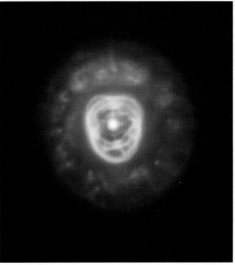

◀ THE ESKIMO Nebula (NGC 2392) is a planetary nebula located in Gemini the Twins. If the sky conditions warrant, use magnifications in excess of 200x for the most detailed view.

PETER AND SUZIE ERICKSON/ADAM BLOCK/NOAO/AURA/NSF

AL AND ANDY FERAVOMI/ADAM BLOCK/NOAO/AURA/NSF

▲ NGC 1931 in Auriga is a small cluster of stars embedded in nebulosity. It lies in a rich area of sky only 1° west of open cluster M36.

MAP 5

9h
50°
8h
7h
↑ Map 1 ↑

URSA MAJOR

NGC 2126

40°

NGC 2537

21

AURIGA

31

LYNX

NGC 2281

NGC 2419

9h
30°

59

UU

65

Map 6 ←

ο

←

π

θ

Castor
α
ρ

χ

χ

σ

τ

NGC 2371/2

β Pollux

ψ

φ

ι

ω

υ

NGC 2331

CANCER

20°

η

κ

NGC 2266

μ

HO 343

GEMINI

ε

M35

θ

ω

NGC 2158

NGC 2420

δ

ζ

μ

1

NGC 2392

η

ζ

BU 1413

ζ

ν

NGC 2174 □

χ²

λ

J900

STT 165

γ

ORION

8h

7h

↓ Map 11 ↓

NORTH EQUATORIAL 2

This map shows objects on early evenings in
autumn and early winter north of the equator, and
in spring and summer south of that line.

**Constellation
boundary**

**Constellation
figure**

Magnitudes

Sirius   • 4.0

0.0   • 5.0

1.0   • 6.0

2.0   • 7.0

3.0   • 8.0

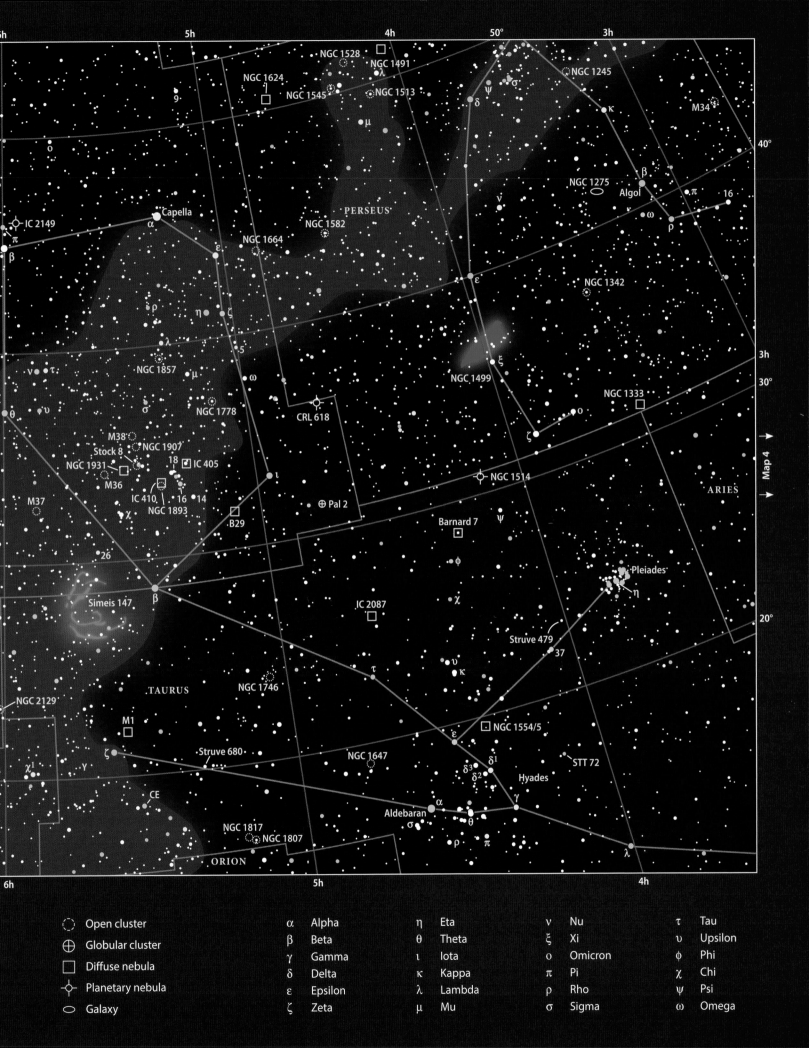

MAP **6** NORTH EQUATORIAL 3

TONY AND DAPHNE HALLAS

▲ **THE SPECTACULAR** magnitude 9.0 spiral galaxy NGC 2903 in Leo is not in Messier's catalog or the Caldwell catalog.

Winter galaxies

The star map on the following pages covers a region visible in the eastern sky on late February evenings north of the equator. It features Cancer the Crab, the northern part of Leo the Lion, Leo Minor the Lion Cub, and the southern portion of Ursa Major the Great Bear. Apart from Cancer's two open star clusters cataloged by Charles Messier, this deep-sky region features galaxies — including some great ones.

Unfortunately, Cancer is a faint constellation. If your sky is even moderately light-polluted, finding it may be a chore. The easiest way is to draw a line from Castor and Pollux, the brightest stars of Gemini, to Regulus, the brightest star of Leo. Cancer lies at the mid-point of this line.

At Cancer's center lies the **Beehive Cluster** (M44). You also may hear this object referred to as the Praesepe. To the naked eye from a dark site, M44 looks like a uniform haze against a starless background. This is not an object for large telescopes. Use 7x50 or 10x50 binoculars or a telescope/eyepiece combination that gives at least a 2° field of view.

Now it's time to observe some galaxies. One thing longtime observers have learned is that, no matter how big your scope is, you'll want a bigger one when you're looking at galaxies. Galaxies aren't that bright, and their light spreads out over a much larger area than, say, a planet. This doesn't mean a small scope won't show anything, just that more details emerge as

you increase aperture. Also, do your best to get away from any light pollution when you observe galaxies. The cores of the brightest galaxies may still show through, but you'll lose any detail in spiral arms or other outer regions. Finally, filters — which allow only a selected portion of light through — don't help. Galaxies contain all types of objects, so the light they emit contains a broad swath of the spectrum. Using a filter to view a galaxy simply will make it appear fainter.

Find Lambda Leonis just west of the Sickle of Leo and move 1½° south to the magnitude 9.0 spiral galaxy **NGC 2903**. A 10-inch telescope shows a halo measuring 4' by 2' surrounding the galaxy's bright nucleus.

At the southern tip of Leo Minor, find **NGC 3344**, a magnitude 9.9 face-on spiral galaxy. A big scope shows the spiral arms well. Through small scopes, you'll see the nucleus surrounded by a faint, mottled area.

Find Mu Ursae Majoris near the border with Leo Minor and move ¾° west to **NGC 3184**. This magnitude 9.8 galaxy is a spiral, but you'll need at least an 8-inch scope to prove that to yourself. Through a smaller scope, you'll see a uniform 4'-wide spot, with only a slight brightening in the center.

For a real galactic treat, look 5½° east of Chi Ursae Majoris for spiral galaxy **M106**. This magnitude 8.3 wonder displays spiral arms with bright blue areas, which are star-forming regions, and vast dust lanes.

HICKSON 44 is a compact galaxy group in Leo comprising elliptical galaxy NGC 3193 (left), tightly wound spiral galaxy NGC 3190 (center), and S-shaped spiral NGC 3187. ADAM BLOCK/NOAO/AURA/NSF

Designation	Right ascension	Declination	Magnitudes	Separation
Zeta Cancri	8h12m	17°39'	5.6, 6.3	5.3"
STF 1224	8h27m	24°32'	7.1, 7.6	5.7"
Phi² Cancri	8h27m	26°56'	6.3, 6.3	5.1"
Iota Ursae Majoris	8h59m	48°02'	3.9, 9.5	4.5"
66 Cancri	9h01m	32°15'	5.9, 8.0	4.5"
75 Cancri	9h09m	26°37'	6.0, 9.1	12.0"
KUI 42	9h42m	31°17'	5.9, 13.6	28.4"
HJ 469	9h45m	18°52'	6.5, 12.6	31.2"
KUI 49	10h19m	46°46'	6.5, 12.2	27.8"
Gamma Leonis	10h20m	19°50'	2.6, 3.8	4.5"
40 Leo Minoris	10h43m	26°19'	5.6, 12.6	18.4"
54 Leonis	10h56m	24°45'	4.5, 6.3	6.5"
51 Ursae Majoris	11h05m	38°14'	6.1, 12.6	8.2"
Nu Ursae Majoris	11h19m	33°05'	3.7, 10.1	7.3"
57 Ursae Majoris	11h29m	39°20'	5.4, 8.4	5.5"
HJ 503	11h36m	27°46'	5.8, 10.2	21.6"
65 Ursae Majoris	11h55m	46°29'	6.7, 8.3	3.7"

TOM BASH AND JOHN FOX/ADAM BLOCK/NOAO/AURA/NSF

THE BEEHIVE Cluster (M44) is an easy naked-eye object from a dark site. Because it's big, use binoculars or a low-power telescope/eyepiece combination to view it.

SPIRAL GALAXY NGC 3486 is a compact object that glows at magnitude 10.3. You'll need a big scope to see detail in its spiral arms. ADAM BLOCK/JEFF HAPEMAN/NOAO/AURA/NSF

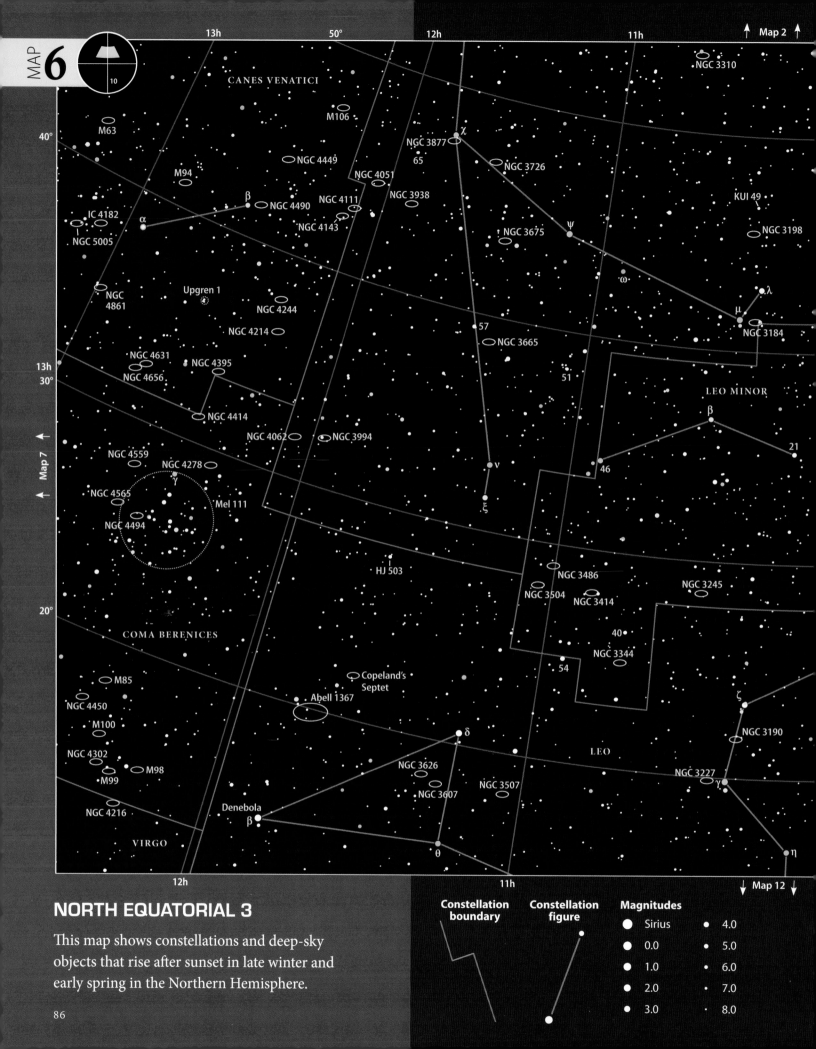

MAP
6
10

13h
50°
12h
11h

↑ Map 2 ↑

NGC 3310

CANES VENATICI

M106

M63

χ
NGC 3877
65

M94
NGC 4449
NGC 3726

β
NGC 4490
NGC 4111
NGC 3938
KUI 49
ψ
NGC 3675
IC 4182
α
NGC 4143
NGC 3198

NGC 5005

ω
λ
μ
NGC 3184

Upgren 1
NGC 4861
NGC 4244
57
NGC 3665

NGC 4214
51

NGC 4631
NGC 4395
LEO MINOR

13h
30°
NGC 4656
β

NGC 4414

21

NGC 4062
NGC 3994
← Map 7
NGC 4559
46
ν

NGC 4278
γ
NGC 4565
Mel 111
ξ
←
NGC 4494

HJ 503
NGC 3486
NGC 3245
NGC 3504
NGC 3414

20°
40
NGC 3344

COMA BERENICES
54

M85
Copeland's
Septet

NGC 4450
Abell 1367
ζ

M100
δ
NGC 3190

NGC 4302
LEO
NGC 3626
γ
M98
NGC 3507
NGC 3227
M99
NGC 3607

NGC 4216
Denebola
β
η

VIRGO
θ

12h
11h
↓ Map 12 ↓

NORTH EQUATORIAL 3

This map shows constellations and deep-sky
objects that rise after sunset in late winter and
early spring in the Northern Hemisphere.

Constellation boundary	Constellation figure	Magnitudes	
		● Sirius	· 4.0
		● 0.0	· 5.0
		● 1.0	· 6.0
		● 2.0	· 7.0
		● 3.0	· 8.0

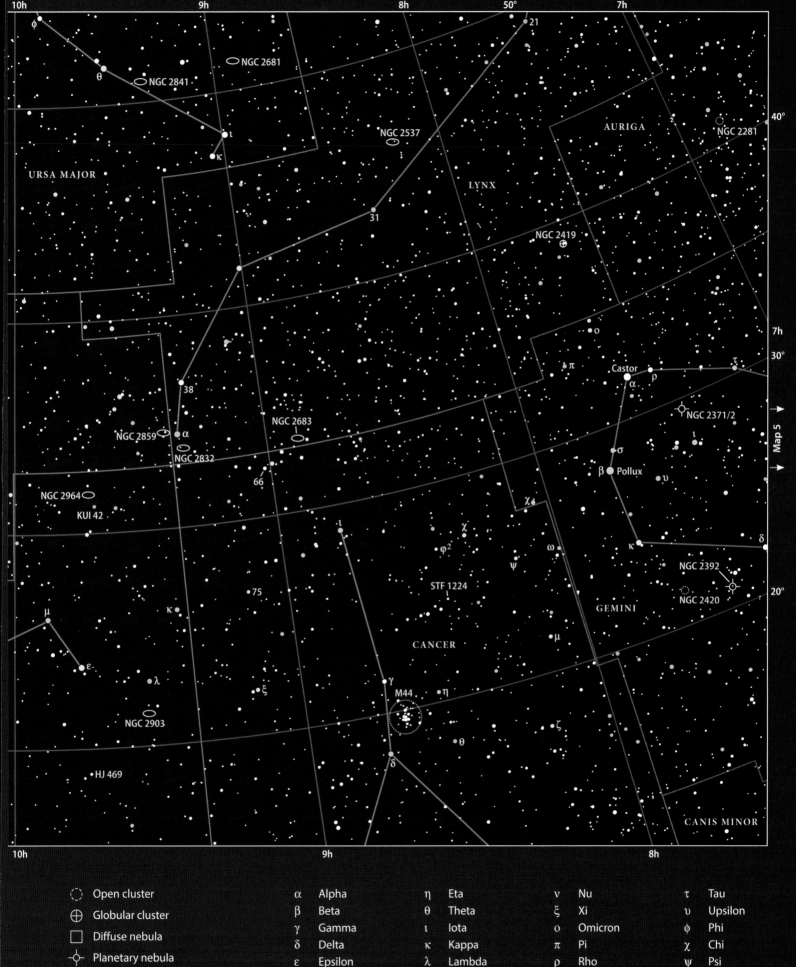

Open cluster

Globular cluster

Diffuse nebula

Planetary nebula

α	Alpha	η	Eta	ν	Nu	τ	Tau
β	Beta	θ	Theta	ξ	Xi	υ	Upsilon
γ	Gamma	ι	Iota	ο	Omicron	φ	Phi
δ	Delta	κ	Kappa	π	Pi	χ	Chi
ε	Epsilon	λ	Lambda	ρ	Rho	ψ	Psi

MAP 7 NORTH EQUATORIAL 4

 NGC 4244 in Canes Venatici is seven times longer than it is wide. As you look toward the nucleus, you won't see much detail beyond a gradual brightening. JOE NAUGHTON AND STEVE STAFFORD/ADAM BLOCK/NOAO/AURA/NSF

Bright galaxies

The next map, like the last, is chock-full of bright galaxies. They mainly reside in the 38th-largest constellation, Canes Venatici, and in Coma Berenices, number 42 on the constellation size list. In contrast, the 13th-largest constellation, Boötes, has far fewer deep-sky objects.

Start by observing two globular clusters. Magnitude 9.2 **NGC 5466** in Boötes appears uniform in brightness. A 12-inch scope reveals two dozen bright stars set against a background glow of unresolvable faint stars.

The other globular is **M3** in Canes Venatici's southern reaches. It lies midway between Cor Caroli (Alpha Canum Venaticorum) and Arcturus (Alpha Boötis). Even small telescopes reveal a lot of detail in this magnitude 6.3 cluster, but point a 12-inch its way, and you'll find a grainy sphere with a bright center and more than 100 stars near its edge.

Also in Canes Venatici, you'll find such standout galaxies as the **Sunflower Galaxy** (M63) and the fabulous **Whirlpool Galaxy** (M51). Point any scope at these objects under a dark sky, and you won't be disappointed. But lesser-known star cities also inhabit this constellation.

Almost 3° north of Beta Canum Venaticorum lies the irregular galaxy **NGC 4449**. This galaxy has quite a different appearance: rectangular. It measures 4' by 2' and shines at magnitude 9.6.

Look 2° southwest of 5th-magnitude 6 Canum Venaticorum for **NGC 4244**. This magnitude 10.4 edge-on spiral galaxy shows disparity between its length and width — it measures 15' by 2'.

Find **NGC 5005** 3° southeast of Alpha Canum Venaticorum. This magnitude 9.8 spiral is an oval twice as long as wide. Most telescopes allow you to pick out an extended central region with a bright nucleus. A nice foreground star shines at 9th magnitude 12' west of NGC 5005.

Coma Berenices contains eight Messier objects, seven of which are galaxies. The exception is globular cluster **M53**, which lies 1° northeast of Alpha Comae Berenices. You can resolve the outer stars of this magnitude

7.7 object, but its central region, although broad, is too concentrated.

The galaxies Messier cataloged (**M64, M85, M88, M91, M98, M99,** and **M100**) are all spirals and worth detailed observing sessions. A fun way to compare these objects is to make small sketches of them on the same sheet of paper. An example of an elliptical galaxy is **NGC 4494**, which lies 3° south-southeast of Gamma Comae Berenices. This magnitude 9.8 galaxy measures 2' across and is ever-so-slightly oval.

Cap off your night viewing objects in Coma with **NGC 4565**, a spectacular magnitude 9.6 edge-on spiral with a small central bulge. Large scopes will reveal a dust lane; it runs the whole length of the galaxy but is easiest to see silhouetted against the core.

GREG ALLEGRETTI

 **THE SUNFLOWER GALAXY** (M63) looks like a disk with a bright center when seen through a small telescope. Larger instruments bring out the arms' spiral structure. M63 shines at magnitude 8.6.

▲ **EDGE-ON SPIRAL GALAXY** NGC 4565 in Coma Berenices shows a small central bulge. Large telescopes will show the dust lane running lengthwise along the galaxy.

▲ **SPIRAL GALAXY** M106 in Canes Venatici offers lots of detail for observers with 8-inch and larger telescopes: a bright, elongated nucleus; dust lanes; and two large spiral arms.

▲ **THE WHIRLPOOL GALAXY** (M51) in Canes Venatici is easy to find because it sits 3½° from Alkaid (Eta Ursae Majoris), which is the end star in the Big Dipper's handle. MIKE GLENNY

▲ **THE WHALE GALAXY** (NGC 4631) is a spiral galaxy that's being distorted by its small companion, elliptical galaxy NGC 4627. The Whale shines at magnitude 9.2.

Designation	Right ascension	Declination	Magnitudes	Separation
2 Canum Venaticorum	12h16m	40°39'	5.9, 8.2	11.5"
11 Comae Berenices	12h21m	17°47'	4.9, 12.9	9.1"
24 Comae Berenices	12h35m	18°22'	5.2, 6.7	20.3"
35 Comae Berenices	12h53m	21°14'	5.1, 9.1	28.7"
Alpha Canum Venaticorum	12h56m	38°18'	2.9, 5.4	19.4"
BU 925	12h57m	43°33'	7.0, 12.5	6.9"
1 Boötis	13h41m	19°57'	5.7, 8.6	4.7"
Tau Boötis	13h47m	17°27'	4.5, 11.1	4.8"
HJ 1244	13h53m	42°11'	7.0, 11.5	6.9"
Kappa Boötis	14h14m	51°47'	4.6, 6.6	13.4"
Pi¹ Boötis	14h41m	16°25'	4.9, 5.8	5.6"
Struve 1884	14h48m	24°22'	6.4, 7.8	2.3"
STT 289	14h56m	32°17'	6.1, 9.6	4.7"
Zeta² Coronae Borealis	15h39m	36°38'	5.1, 6.0	6.3"

▲ **THROUGH A SMALL TELESCOPE,** M94 in Canes Venatici looks like an elliptical galaxy. Actually, it's a bright (magnitude 8.2) spiral galaxy with tightly wound arms.

▲ **SPIRAL GALAXY** NGC 5371 in Canes Venatici has a bright core and faint arms. The nearby bright star is magnitude 8.7 SAO 44805. It lies 2' from NGC 5371's center.

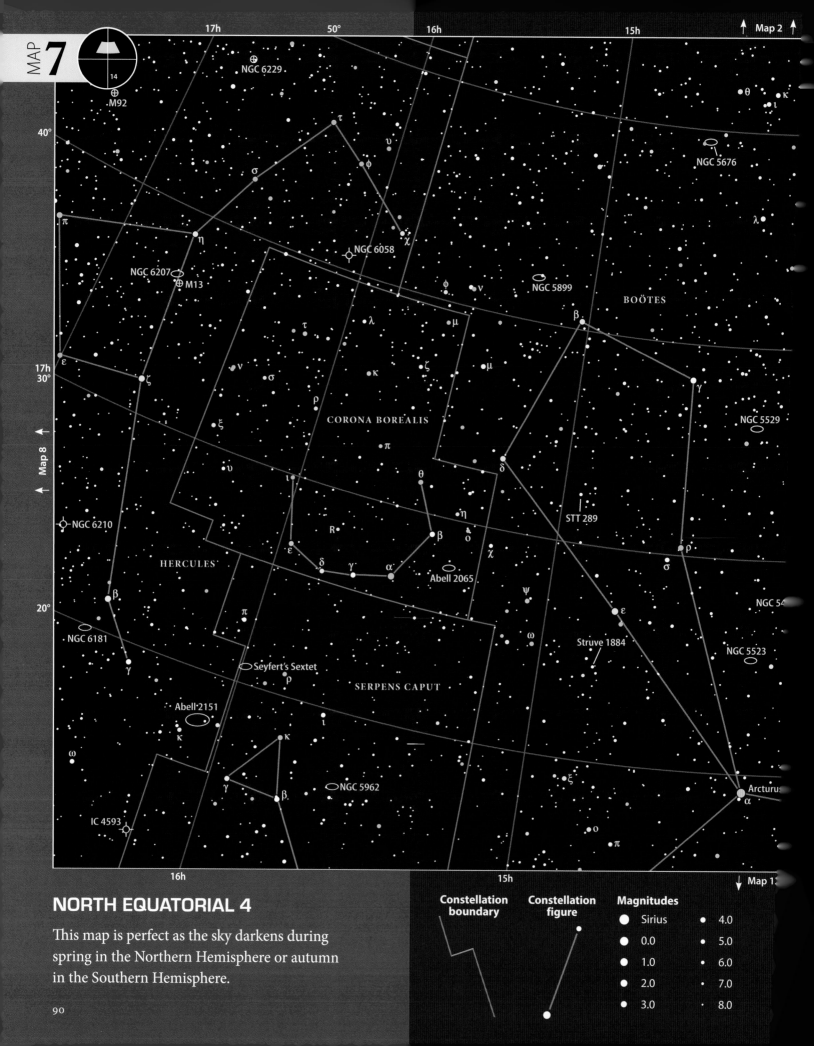

MAP
7
14

17h

50°

16h

15h

↑ Map 2 ↑

NGC 6229

M92

τ

υ

σ

φ

θ

κ

ι

40°

χ

NGC 5676

π

η

NGC 6058

λ

BOÖTES

NGC 6207

M13

φ

ν

NGC 5899

β

ε

τ

μ

17h
30°

ζ

ν

σ

κ

ζ

μ

γ

ρ

CORONA BOREALIS

NGC 5529

Map 8

ξ

π

δ

STT 289

υ

θ

NGC 6210

ι

δ

ρ

HERCULES

ε

R

β

η
&
ο

σ

γ

α

χ

β

Abell 2065

ψ

20°

π

ε

NGC 6181

ω

Struve 1884

NGC 5523

γ

Seyfert's Sextet

ρ

SERPENS CAPUT

Abell 2151

κ

ι

ξ

ω

κ

γ

β

NGC 5962

Arcturus

IC 4593

α

ο

π

16h

15h

↓ Map 1

NORTH EQUATORIAL 4

This map is perfect as the sky darkens during spring in the Northern Hemisphere or autumn in the Southern Hemisphere.

Constellation boundary	Constellation figure	Magnitudes	
		● Sirius	● 4.0
		● 0.0	· 5.0
		● 1.0	· 6.0
		● 2.0	· 7.0
		● 3.0	· 8.0

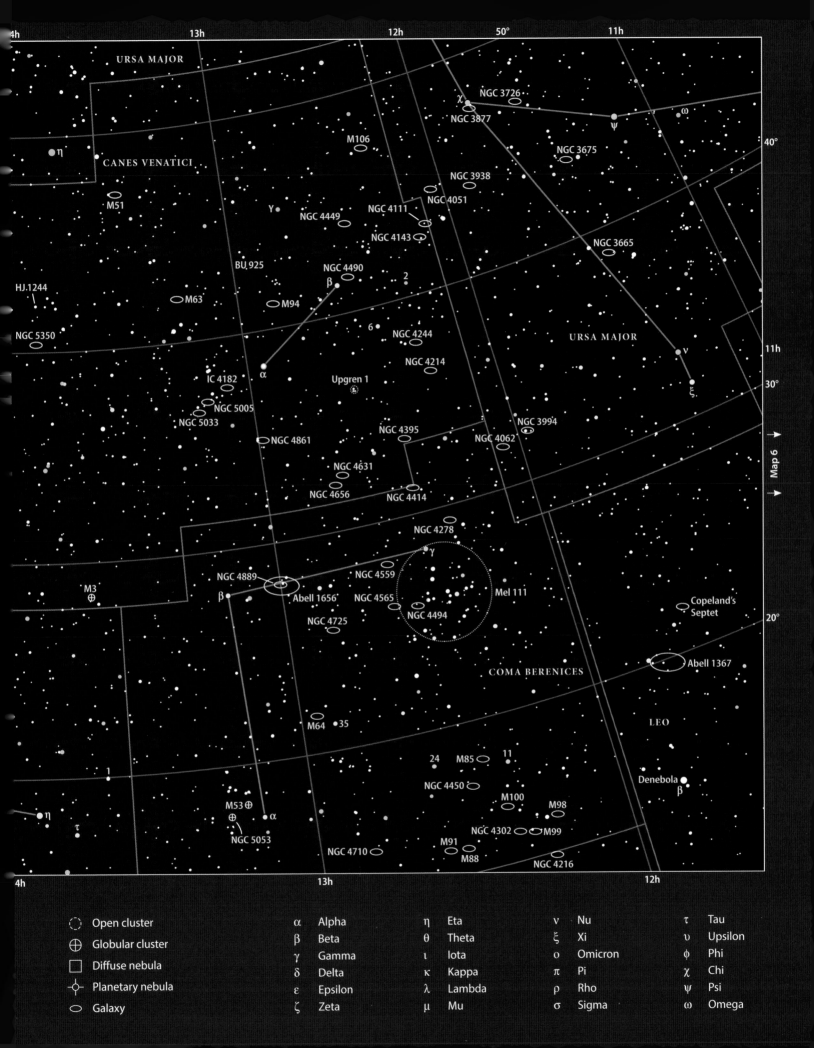

Star chart labels:

Top row (RA/Dec grid): 14h · 13h · 12h · 50° · 11h · 40°

URSA MAJOR

CANES VENATICI

η

M51

NGC 3726
χ
NGC 3877
ψ
ω

M106
NGC 3675

NGC 3938
NGC 3665
NGC 4051
NGC 4111
NGC 4143

γ
NGC 4449

BU 925
NGC 4490
β
2

HJ.1244
M63
M94
6
NGC 4244

NGC 5350
α
NGC 4214
URSA MAJOR
ν
11h
30°

IC 4182
Upgren 1
ξ

NGC 5005
NGC 3994
NGC 5033
NGC 4861
NGC 4395
NGC 4062

NGC 4631
Map 6
NGC 4656
NGC 4414

NGC 4278
γ
NGC 4889
NGC 4559
Mel 111
β
Abell 1656
NGC 4565
NGC 4494
M3
NGC 4725
Copeland's Septet
20°

Abell 1367

COMA BERENICES

LEO

M64
35

Denebola
β

24
M85
11

NGC 4450
M100
M98
1
η
τ
M53
α
NGC 4302
M99
NGC 5053
M91
NGC 4710
M88
NGC 4216

Bottom row: 14h · 13h · 12h

Legend:

- ⟡ Open cluster
- ⊕ Globular cluster
- ☐ Diffuse nebula
- ⬥ Planetary nebula
- ⬭ Galaxy

α	Alpha	η	Eta	ν	Nu	τ	Tau
β	Beta	θ	Theta	ξ	Xi	υ	Upsilon
γ	Gamma	ι	Iota	ο	Omicron	φ	Phi
δ	Delta	κ	Kappa	π	Pi	χ	Chi
ε	Epsilon	λ	Lambda	ρ	Rho	ψ	Psi
ζ	Zeta	μ	Mu	σ	Sigma	ω	Omega

MAP **8** **NORTH EQUATORIAL 5**

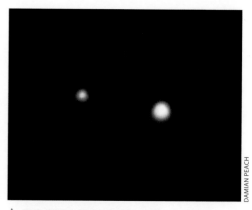

▲ **THE SKY'S GREATEST** double star may be Albireo (Beta Cygni), which marks the Swan's head. The gold and sapphire-blue components easily separate in even the smallest telescope.

▲ **NGC 6210** is a planetary nebula in Hercules worth looking at with a low-power eyepiece. At magnifications around 50x, you won't see detail, but its color and uniform brightness will stand out.

▲ **THE RING NEBULA (M57)** in Lyra may be the sky's most-observed planetary nebula. This magnitude 8.8 object measures 1.4' by 1'. TONY HALLAS

A showpiece globular

The constellations Hercules the Hero and Lyra the Harp are the two dominant star patterns on the following map. Denser regions of the Milky Way congregate in our galaxy's arms, preventing us from seeing much of what lies beyond. Instead, nearby star clusters and nebulae abound.

Starting in Hercules, find Pi, Eta, Zeta, and Epsilon Herculis — the Keystone of Hercules. Two-thirds of the way from Zeta to Eta, you'll find **M13**, the eighth-brightest globular cluster in the sky. You can glimpse this magnitude 5.8 cluster without optical aid even from less-than-ideal locations. Any telescope pointed at M13 reveals a glorious swarm of stars. Through an 8-inch scope, you'll see more than 100. Increase the magnification beyond 150x, and try to spot the "propeller," the intersection of three dark lanes near the cluster's center.

Only slightly fainter than M13 is another globular cluster — magnitude 6.5 **M92**. This object lies on a nearly exact line between Iota and Eta Herculis. It's 5° from Iota and 7.6° from Eta.

Less than ½° northeast of M13 lies the magnitude 11.6 spiral galaxy **NGC 6207**. Insert an eyepiece that will give you a bit more than a half-degree field of view, and enjoy NGC 6207 and M13 at the same time.

Before you leave Hercules, check out **NGC 6210**, a small (25" by 15" in diameter) but bright (magnitude 8.8) planetary nebula. At magnifications under 100x, this object's color and uniform brightness are striking.

Midway between Alpha Aquilae and Beta Cygni lies Sagitta the Arrow. This star pattern ranks 86th out of the 88 constellations in size. It does contain a Messier object, however — **M71**. Glowing at 8th magnitude, M71 appears loosely arranged for a globular cluster. An 8-inch telescope resolves several dozen of the brightest stars set against a hazy background.

Because of its small size, parallelogram shape, and zero-magnitude star, Lyra the Harp is one of the easiest constellations to recognize. Two showpiece deep-sky objects reside there.

The **Double Double** (Epsilon Lyrae) is, as the name promises, a pair of double stars. Epsilon[1] is the northern and wider pair. The stars shine at magnitudes 5.0 and 6.1, and their separation is 2.6". The components of Epsilon[2] lie 2.3" from each other with each star glowing at magnitude 5.5. Even a small telescope at 100x will split both pairs.

The **Ring Nebula** (M57) is Lyra's other spectacle. A small scope reveals this object's doughnut shape. Move to a larger scope, and you'll notice the ring isn't exactly round but elongated roughly east-west. The darkened central area is easy to see, but the central star is a challenging object best left to telescopes 16 inches in aperture and larger.

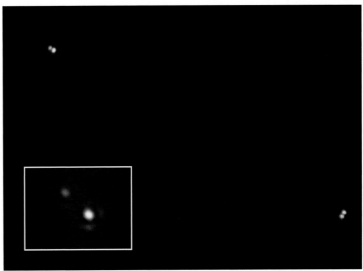

▲ **THE DOUBLE DOUBLE** (Epsilon Lyrae) offers a rewarding pair of targets. Use low power to see both sets in the same field of view. The inset image shows Epsilon² Lyrae. DAMIAN PEACH

◀ **THE HERCULES CLUSTER** (M13) is a beautiful small-telescope target. At magnitude 5.8, it's a naked-eye object when viewed from a dark site. KEES SCHERER

▲ **M92 IN HERCULES** is that constellation's "forgotten" globular cluster because so many observers concentrate their time on M13. Do yourself a favor, however, and don't overlook this magnitude 6.5 gem.

/// DOUBLE-STAR DELIGHTS — MAP 8

Designation	Right ascension	Declination	Magnitudes	Separation
Kappa Herculis	16h08m	17°03'	5.3, 6.5	28.1"
Omega Herculis	16h25m	14°02'	4.5, 11.0	28.4"
52 Herculis	16h49m	45°59'	4.9, 10.4	1.8"
HO 412	17h08m	35°56'	5.4, 11.4	20.0"
Delta Herculis	17h15m	24°50'	3.2, 8.3	9.5"
Rho Herculis	17h24m	37°08'	4.5, 5.5	4.1"
95 Herculis	18h01m	21°35'	5.1, 5.2	6.3"
100 Herculis	18h08m	26°05'	5.9, 6.0	14.2"
102 Herculis	18h09m	20°49'	4.3, 11.8	23.4"
Nu² Lyrae	18h50m	32°33'	5.2, 12.7	19.0"
Gamma Lyrae	18h59m	32°41'	3.3, 12.1	13.4"
STF 2486	19h12m	49°50'	6.6, 6.8	7.9"
Eta Lyrae	19h14m	39°12'	4.0, 8.0	28.0"
Theta Cygni	19h37m	50°12'	4.6, 12.9	3.2"
17 Cygni	19h46m	33°44'	5.0, 9.2	26.0"
Zeta Sagittae	19h49m	19°06'	5.5, 9.0	8.0"
Eta Cygni	19h56m	35°05'	4.0, 12.0	7.4"
Theta Sagittae	20h10m	20°54'	6.4, 8.9	12.0"

CHRIS SCHUR

MAP **8**
18

NORTH EQUATORIAL 5

Use this map on early summer evenings in
the Northern Hemisphere and on early winter
evenings south of the equator.

Constellation boundary **Constellation figure** **Magnitudes**

	Magnitudes	
◯ Sirius		· 4.0
⬤ 0.0		· 5.0
● 1.0		· 6.0
● 2.0		· 7
● 3.0		· 8

Labels on map:

NGC 7000, NGC 7027, IC 5067, Sh 2-112, Deneb, α, ξ, ν, o², o¹, NGC 6826, ι, θ, STF 2486, NGC 6811, NGC 6742, NGC 6914, NGC 6866, δ, 40°, σ, τ, NGC 6910, IC 1318, γ, NGC 6742, PK 80-6.1, CYGNUS, λ, M29, NGC 6888, NGC 6819, NGC 6745, ε, Veil Nebula, NGC 6871, Sh 2-101, NGC 6857, 17, χ, η, NGC 6791, η, θ, Vega, μ, α, δ, ζ, τ, κ, ι, LYRA, 21h 30°, NGC 6940, NGC 6894, NGC 6834, φ, γ, λ, M57, ν, NGC 6765, M56, NGC 6885, VULPECULA, NGC 6882, M1-92, Albireo, β, NGC 6800, 13, 20°, NGC 6830, NGC 6823, α, M27, NGC 6905, θ, NGC 6820, DELPHINUS, NGC 6886, η, γ, M71, ζ, NGC 6802, Cr 399, H20, δ, α, β, Pal 10, ε, SAGITTA, NGC 6891, AQUILA, 1, Map 9

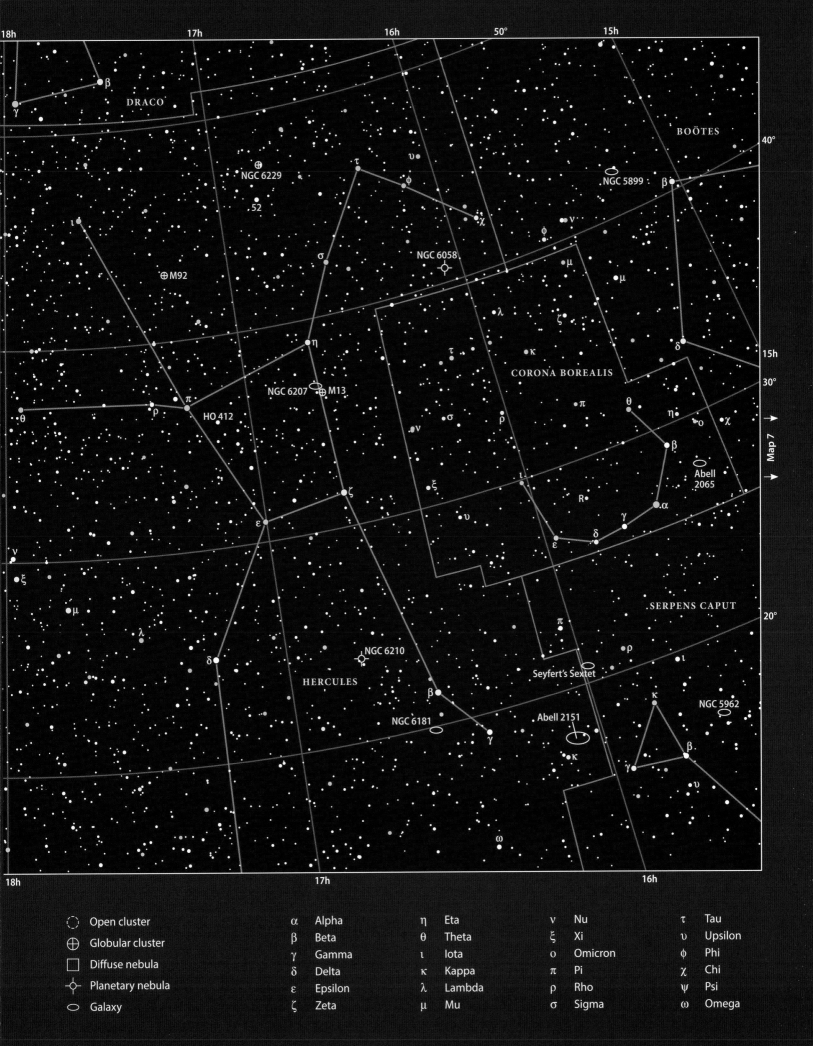

Open cluster

Globular cluster

Diffuse nebula

Planetary nebula

Galaxy

α	Alpha	η	Eta	ν	Nu	τ	Tau
β	Beta	θ	Theta	ξ	Xi	υ	Upsilon
γ	Gamma	ι	Iota	o	Omicron	φ	Phi
δ	Delta	κ	Kappa	π	Pi	χ	Chi
ε	Epsilon	λ	Lambda	ρ	Rho	ψ	Psi
ζ	Zeta	μ	Mu	σ	Sigma	ω	Omega

MAP **9**

NORTH EQUATORIAL 6

KEES SCHERER

▲ **THE EASTERN VEIL NEBULA** is part of a vast cloud of dust and gas ejected by a massive star in its violent death throes several thousand years ago.

Flying with the Swan

The next map shows the stars corresponding to autumn in the Northern Hemisphere. The eastern regions of Cygnus the Swan and the northern part of Pegasus the Winged Horse dominate the chart. Cygnus lies along the Milky Way and contains a vast number of clusters and nebulae.

Start 3° east of Deneb (Alpha Cygni) and locate the **North America Nebula** (NGC 7000). Many observers can see this object's outline even without binoculars, but others require help. It's big (2° by 1⅔°), so use binoculars or a wide-field view through a telescope to attack it. If you're still having trouble, use either a nebula filter or a more restrictive OIII filter. Some observers report positive results by hand-holding the filter and not using a telescope at all.

If you have an OIII filter, observe the **Veil Nebula** (NGC 6992/5). The Veil is the remnant of a supernova that exploded less than 10,000 years ago. It lies some 2,100 light-years away and measures an immense 2.7° by 3.8°. Approach this object in two ways. You can try to capture all of it in one view, but you'll need an eyepiece/telescope combination yielding a 4° field. While you'll capture it, you won't see much detail. To get a clearer view, use an eyepiece (plus OIII filter) in the 50x-to-75x range and scan the Veil by moving your scope to follow its full extent.

On Map 3, about 1½° east-northeast of Theta Cygni lies the **Blinking Planetary** (NGC 6826). Through small scopes, you'll see the central star if you look at NGC 6826 using direct vision. Look a bit away from the

KEES SCHERER

▲ **THE NORTH AMERICA NEBULA** (NGC 7000) in Cygnus measures 2° across. Can you see this object, which lies near Deneb (Alpha Cygni), with your unaided eyes? If not, try looking through a nebula filter.

central star with "averted vision," and it will be swallowed in the planetary's nebulosity. So, look directly at — and then away from — the star to "blink" the planetary. Use a small scope to try this because telescopes larger than 6 inches keep the central star visible at all times.

Fall in the Northern Hemisphere often is called "planetary season" because of the large number of planetary nebulae visible. Magnitude 10.7 **NGC 7008** (also on Map 3) lies in a region devoid of bright stars. NGC 7008 lies almost 10° north of Deneb and measures 83" across.

For a new perspective on planetary nebulae, turn your scope toward the **Dumbbell Nebula** (M27) in Vulpecula. M27 is one of the largest and

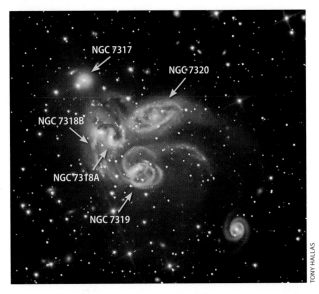

NGC 7317
NGC 7320
NGC 7318B
NGC 7318A
NGC 7319

TONY HALLAS

◀ **STEPHAN'S QUINTET** is a group of five galaxies in Pegasus. You can spot all five through a 12-inch telescope, but to see details in these objects, you'll need to use a 20-inch or larger scope. At magnitude 12.5, NGC 7320 shines brightest.

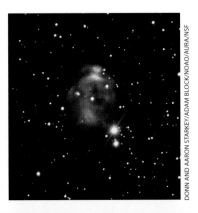

DONN AND AARON STARKEY/ADAM BLOCK/NOAO/AURA/NSF

▶ **PLANETARY NEBULA** NGC 7008 (Map 3) in Cygnus shines at magnitude 10.7. Through a medium-sized telescope, this object looks like an irregular broken ring. You'll need a 12-inch or larger scope to see any color, however.

/// DOUBLE-STAR DELIGHTS — MAP 9

Designation	Right ascension	Declination	Magnitudes	Separation
Gamma Delphini	20h47m	16°08'	4.5, 5.5	9.6"
Kappa Pegasi	21h45m	25°36'	4.5, 10.0	14.0"
STF 2841	21h54m	19°43'	6.5, 8.0	22.3"
Iota Pegasi	22h07m	25°21'	4.0, 11.4	3.7"
STF 2894	22h19m	37°46'	6.2, 8.4	15.6"
STF 2906	22h27m	37°26'	6.4, 10.0	4.4"
8 Lacertae	22h36m	39°38'	6.6, 6.8	22.4"
13 Lacertae	22h44m	41°48'	5.2, 10.6	14.6"
15 Lacertae	22h52m	43°17'	5.2, 12.2	25.7"
Beta Pegasi	23h04m	28°04'	2.6, 11.8	8.5"
75 Pegasi	23h38m	18°24'	5.4, 11.6	27.7"

LOUS AND JENNIFER GOLDRING/ADAM BLOCK/NOAO/AURA/NSF

▲ **SPIRAL GALAXY** NGC 7331 in Pegasus is one of the brightest galaxies not in Messier's catalog. It shines at magnitude 9.5. Some half dozen galaxies accompany NGC 7331.

brightest (magnitude 7.3) planetaries in the sky, mostly because of its distance, a scant 800 light-years. Binoculars reveal this object, and a small telescope shows its lobes (the ends of the dumbbell). Through a 12-inch or larger scope, you'll see irregularities in M27's surface and thin, faint arcs at both lobe ends. More than a dozen faint stars lie superimposed over the Dumbbell, which measures about 6' across.

As we've progressed through the past few star maps, we've also moved farther from the Milky Way, so more galaxies are visible. Start with magnitude 9.5 spiral galaxy **NGC 7331**, which lies 4.3° north-northwest of Eta Pegasi. This galaxy is part of a group that numbers less than a dozen. Use a low-power eyepiece through a 10-inch scope to see several other members. The galaxy is 10.5' long and a third as wide. The bright, elongated central region washes out the fainter spiral arms through telescopes less than 12 inches in aperture.

Finally, if you have access to a large scope, train it on **Stephan's Quintet** (NGC 7317, NGC 7318A, NGC 7318B, NGC 7319, and NGC 7320), a collection of galaxies ½° southwest of NGC 7331. In the quintet, NGC 7320 is the brightest, glowing meekly at magnitude 12.5. The faintest, NGC 7317, shines 13 times fainter at magnitude 15.3.

▲ **THE DUMBBELL NEBULA** (M27) in Vulpecula the Fox is one of the largest and closest planetary nebulae. You'll see this magnitude 7.3 object easily through small telescopes (or even binoculars), but it appears spectacular through large ones. MIKE DURKIN

MAP **9**
22
↑ Map 3 ↑

1h
50°
0h
23h

o
NGC 185
π
NGC 147

CASSIOPEIA

β

α NGC 7243

4

LACERTA

5

2

NGC 7209

40°

M31
ν
NGC 205
μ
M32

ψ
λ

κ
ι

NGC 7662

NGC 404
β

And I

15

6

13

θ
ρ
σ

ANDROMEDA

o

NGC 7640

BL Lac

π

8

1h
30°

δ

ε

STF 2894

STF 2906

1

Map 4 ←

α

NGC 7331
Stephan's Quintet

NGC 1

Jones 1

π

NGC 7457
η

NGC 7217

20°

ψ

NGC 7741

β

o

PEGASUS

υ
τ

μ

NGC 7332

λ

χ

NGC 7678

ι

PISCES

φ

75

NGC 7814

γ

Pegasus Dwarf

α

0h
23h
↓ Map 15 ↓

NORTH EQUATORIAL 6

View the objects on this map in the early evening
during late summer in the Northern Hemisphere
and late winter south of the equator.

Constellation boundary	Constellation figure	Magnitudes	
		Sirius	4.0
		0.0	5.0
		1.0	6.0
		2.0	7.0
		3.0	8.0

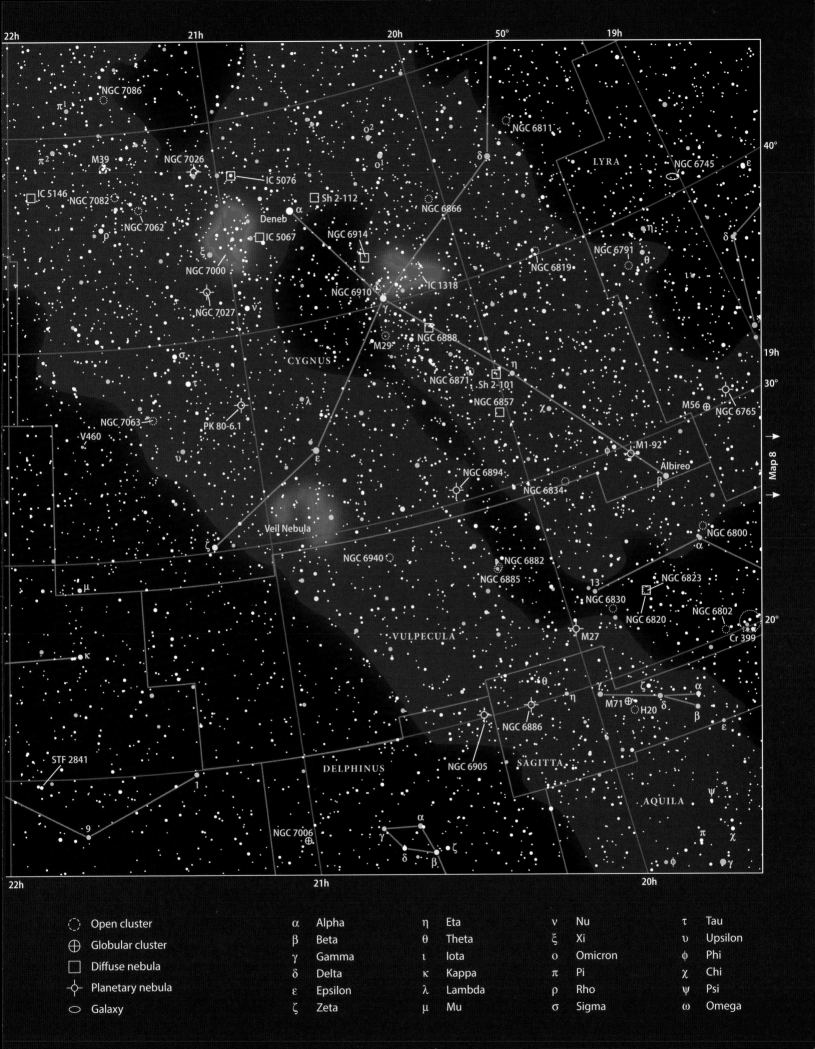

NGC 7086

π¹

π²
M39   NGC 7026   IC 5076

IC 5146   NGC 7082   Sh 2-112

NGC 7062   Deneb   α   NGC 6914

ρ   ξ   IC 5067

NGC 7000   NGC 6910   IC 1318

NGC 7027   ν   γ

σ   M29   NGC 6888

CYGNUS

τ   η   NGC 6871   Sh 2-101

λ   NGC 6857

NGC 7063   PK 80-6.1   χ

V460

υ   ε

Veil Nebula   NGC 6894   φ   M1-92

ζ   NGC 6834   Albireo   β

μ   NGC 6940   NGC 6882   α

NGC 6885   13   NGC 6823

κ   VULPECULA   M27   NGC 6830   NGC 6802

NGC 6820   Cr 399

STF 2841   θ   γ   ζ   α

η   M71   δ   β

ι   NGC 6886   H20   ε

9   DELPHINUS   NGC 6905   SAGITTA

NGC 7006   AQUILA

α   ψ

γ   π   χ

δ   β   ζ   φ   γ

NGC 6811

LYRA   NGC 6745   ε

η   δ

NGC 6791   θ   ι

NGC 6819   NGC 6866

NGC 6745

M56   NGC 6765

NGC 6800

40°

19h

30°

Map 8

20°

22h   21h   20h

Open cluster						
⊕ Globular cluster	α	Alpha	η	Eta	ν	Nu	τ	Tau
☐ Diffuse nebula	β	Beta	θ	Theta	ξ	Xi	υ	Upsilon
Planetary nebula	γ	Gamma	ι	Iota	ο	Omicron	φ	Phi
Galaxy	δ	Delta	κ	Kappa	π	Pi	χ	Chi
	ε	Epsilon	λ	Lambda	ρ	Rho	ψ	Psi
	ζ	Zeta	μ	Mu	σ	Sigma	ω	Omega

MAP **10** EQUATORIAL REGION 1

▲ **SPIRAL GALAXY** M74 in Pisces is the prototypical face-on spiral. It glows at magnitude 8.5 and lies 35 million light-years away. MARC VAN NORDEN

An ocean of galaxies

Two large but faint constellations dominate the first equatorial star chart — Cetus the Whale and Pisces the Fish. Lots of galaxies populate this region, which lies far from the Milky Way. Charles Messier pinpointed only one object in each of these giant constellations, but a patient observer equipped with an 8-inch or larger telescope under a dark sky will see much more than two bright galaxies.

We'll start with an object that's not a galaxy — planetary nebula **NGC 246**. To find it, make an equal-sided triangle with Phi[1] and Phi[2] Ceti. You'll see the nebula through small telescopes, but this magnitude 10.9 object's details start to emerge only through 6-inch or larger instruments. Double the aperture to 12 inches, and you'll see the planetary's bright northeast rim and five superimposed stars. Add a nebula filter, and NGC 246's irregular inner structure will jump out at you.

Follow NGC 246 with — what else? — **NGC 247** (9° south, on Map 16). The given brightness of this galaxy, magnitude 8.8, is deceptive. Because NGC 247 measures 20' by 7', its light spreads over a larger area than most galaxies'. Its low surface brightness makes it a target for larger scopes, although you can glimpse it through a 3-inch under a dark sky.

With a 12-inch or larger telescope at a dark site, you can spot **IC 1613**, which lies about 1° north-northeast of 26 Ceti. IC 1613 is another low-surface-brightness galaxy, but it's worth finding because it belongs to the Local Group. Use a wide-field eyepiece under a dark sky to scan for this magnitude 9.2 irregular galaxy. Look for a fairly large (20' across) brightening of the sky background.

The showpiece object in Cetus is its sole Messier object: **M77**. This magnitude 8.9 spiral galaxy lies less than 1° east-southeast of Delta Ceti. Through a 4-inch telescope, the core of M77 appears starlike with a wispy halo. Because the halo is denser to the northwest, the galaxy has a vague comet-like appearance. An 8-inch or larger scope reveals structure in the central 2' of this 8' by 7' galaxy, but you'll need at least 12 inches of aperture to resolve the closely wound spiral arms. A magnitude 10 star glows 1.5' to the east-southeast.

Almost exactly 0.5° north-northwest of M77 lies the equally beautiful edge-on spiral **NGC 1055**. You'll see both galaxies in any eyepiece that shows you the entire Full Moon, and they're quite a sight through a large telescope. Through a 6-inch scope, you'll see NGC 1055 as a 4' by 1' spindle aligned east-west. Increase your aperture to 10 inches, and you'll notice the dust lane that cuts this galaxy in two lengthwise. The dark lane's northern edge is easiest to spot.

Hop north to Aries, and a bit more than 1° east of Gamma Arietis (a great double star) to find spiral galaxy **NGC 772**. This magnitude 10.3 object is more than 50 percent longer than it is wide (7.3' by 4.6'). Astronomers classify NGC 772 as a peculiar galaxy, mainly due to its gravitational interaction with NGC 770. You easily can see the nucleus of NGC 770 just off the southern edge of NGC 772's halo.

Aries also is home to a small galaxy group. The brightest member is **NGC 877**, a magnitude 11.9 barred spiral. Use at least a 12-inch telescope to see it and its fainter companions: NGC 876 lies to the southwest, and NGC 871, the brightest of several dim galaxies, is east of NGC 877.

In Eridanus the River, **NGC 1300** deserves your attention. This barred spiral galaxy lies 2° north of Tau[4] Eridani and glows at magnitude 10.4. The galaxy's nucleus stands out nicely. Next, in order of visual prominence, are the two nodules at the ends of the bar. Most difficult to see are NGC 1300's spiral arms, which wind tightly around the inner parts.

Look a bit more than 1.5° southwest of Pi Eridani for peculiar galaxy **NGC 1421**. Barely visible through a 6-inch telescope, NGC 1421 shows lots of detail through a 10-inch or larger scope. It appears spindle-like (3' by 0.5') with its long axis aligned north-south. Crank up the power, and you'll see irregular structure in the form of several knotty regions. The core, which stretches 1', appears brighter on its northern end.

Moving to Pisces, look slightly more than 2° southwest of Mu Piscium for spiral galaxy **NGC 488**. The outer regions of this magnitude 10.3 object are difficult to see, but the central 1' is bright and gets brighter toward the core. NGC 488 measures 5.4' by 3.9'.

Spiral galaxy **NGC 676** shines at 10th magnitude, or maybe 11th. The

▲ **M77 (UPPER LEFT)** and NGC 1055 exemplify face-on and edge-on spiral galaxies, respectively. Both lie about 70 million light-years away in Cetus.

DANIEL VERSCHATSE

JEFF CREMER/ADAM BLOCK/NOAO/AURA/NSF

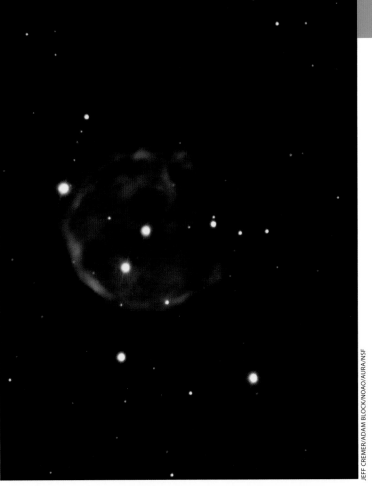

▲ **PLANETARY NEBULA** NGC 246 in Cetus spans nearly 4' and shows irregular texture in its outer ring. Use an OIII filter for the best view.

▲ **SPIRAL GALAXY** NGC 772 in Aries may be interacting with NGC 770, the small elliptical galaxy above it. ADAM BLOCK/NOAO/AURA/NSF

uncertainty comes from a magnitude 9.5 star positioned in front of the galaxy's nucleus. The star is bright enough that you'll have trouble seeing the galaxy through a 6-inch telescope. Through a 12-inch, however, you'll see NGC 676 as lens-shaped, measuring 3' by 1' and aligned roughly north-south. Look for NGC 676 roughly 2° east-northeast of Nu Piscium.

The great face-on spiral galaxy **M74** is Pisces' highlight object. Find it 1.3° east-northeast of Eta Piscium. M74 extends 10' in diameter, so its magnitude 8.5 brightness spreads out a lot. Even a 2.4-inch telescope shows half the galaxy's extent, but a large scope (12 inches and above) reveals both stellar associations and gas clouds. The core is broad — 2.5' across — and condensed. Adding to M74's appearance, six stars lie in our line of sight to this galaxy, two of them superimposed on its nucleus.

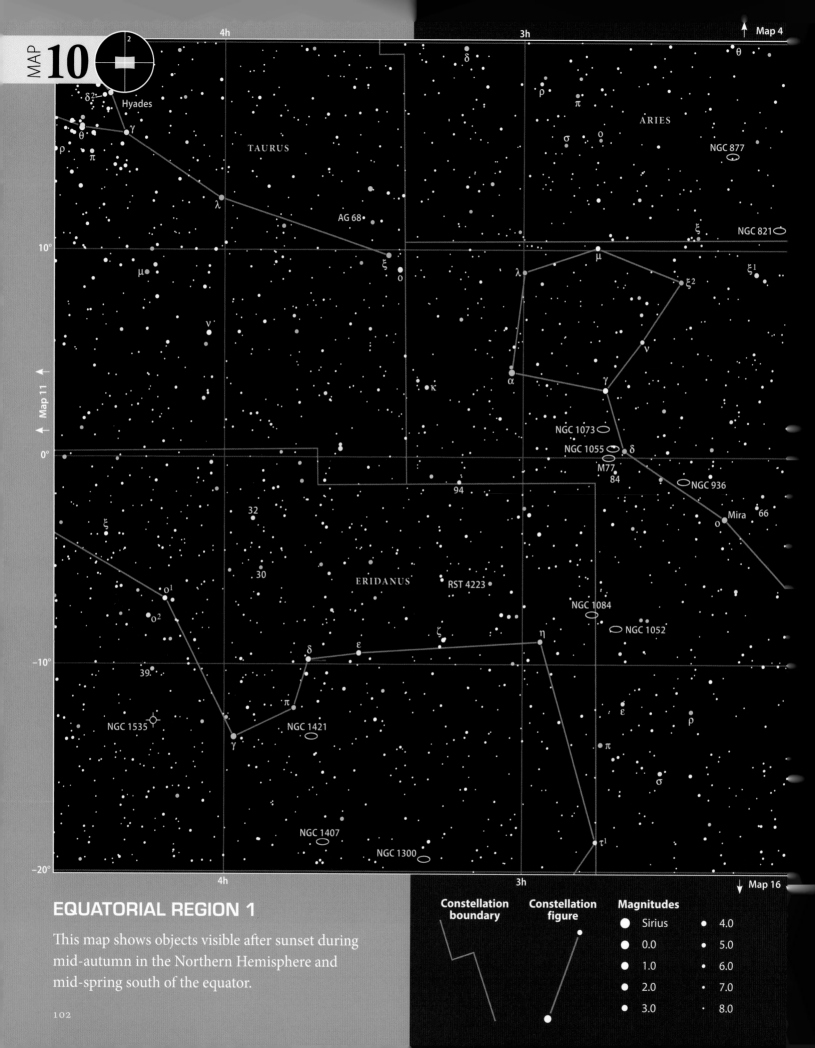

MAP
10
2

Hyades

δ²

θ

ρ

π

γ

λ

TAURUS

AG 68

μ

ν

ξ

o

10°

ARIES

δ

ρ

π

σ

o

NGC 877

ξ

NGC 821

μ

λ

ξ¹

ξ²

ν

κ

α

γ

0°

NGC 1073

NGC 1055

δ

M77

84

NGC 936

Mira

66

o

32

30

ξ

o¹

o²

ERIDANUS

RST 4223

94

NGC 1084

NGC 1052

39

ζ

η

-10°

NGC 1535

δ

ε

ε

π

ρ

NGC 1421

π

γ

σ

NGC 1407

τ¹

NGC 1300

-20°

4h

3h

↑ Map 11

← Map 11

↓ Map 16

EQUATORIAL REGION 1

This map shows objects visible after sunset during
mid-autumn in the Northern Hemisphere and
mid-spring south of the equator.

Constellation boundary

Constellation figure

Magnitudes

Sirius
0.0
1.0
2.0
3.0

4.0
5.0
6.0
7.0
8.0

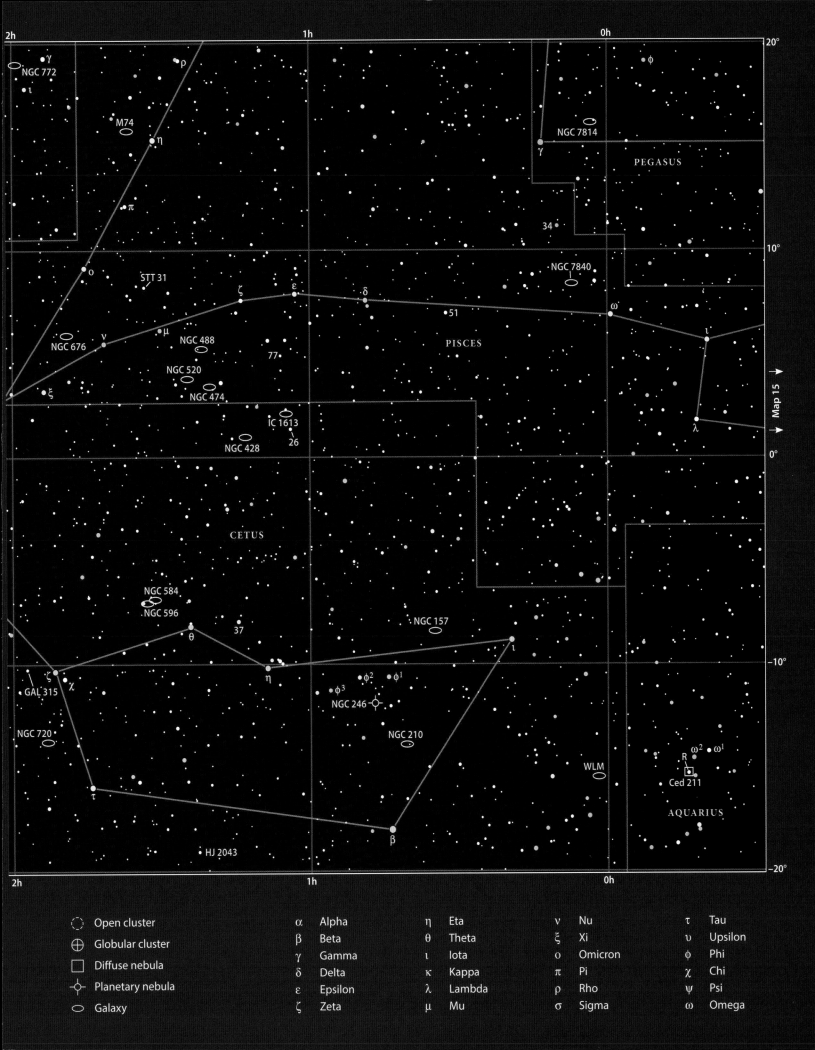

2h   1h   0h

20°

γ
NGC 772
ι
ρ
M74
η
π
φ
NGC 7814
γ
PEGASUS
34
10°
ο
STT 31
ζ
ε
δ
NGC 7840
51
ω
Map 15
ν
μ
NGC 676
NGC 488
ξ
77
NGC 520
PISCES
NGC 474
ι
IC 1613
λ
NGC 428
26
0°
CETUS
NGC 584
NGC 596
θ
37
NGC 157
ι
η
-10°
ζ
χ
φ²
φ¹
GAL 315
φ³
NGC 246
NGC 720
NGC 210
WLM
R
ω²
ω¹
Ced 211
τ
AQUARIUS
HJ 2043
β
-20°

2h   1h   0h

⬡ Open cluster	α Alpha	η Eta	ν Nu	τ Tau
⊕ Globular cluster	β Beta	θ Theta	ξ Xi	υ Upsilon
▢ Diffuse nebula	γ Gamma	ι Iota	ο Omicron	φ Phi
⊕ Planetary nebula	δ Delta	κ Kappa	π Pi	χ Chi
⬭ Galaxy	ε Epsilon	λ Lambda	ρ Rho	ψ Psi
	ζ Zeta	μ Mu	σ Sigma	ω Omega

RYAN STEINBERG AND FAMILY/ADAM BLOCK/NOAO/AURA/NSF

▲ **THE ORION NEBULA** (M42) in Orion's sword is one of the sky's brightest nebulae. Explore it with all your eyepieces and nebula filters.

Nights of the Hunter

The heart of the winter sky fills the next star map. There, we encounter Orion the Hunter, the favorite constellation of many observers. Orion is just the start, however. Taurus, Gemini, Eridanus, Canis Minor, and Canis Major all contribute bright stars. And don't overlook Monoceros the Unicorn. Although it contains no bright stars, Monoceros is a treasure trove of deep-sky objects and double stars.

In Taurus, open cluster **NGC 1817** glows at magnitude 7.7 and spans 16'. Simply put, look through bigger telescopes to see more stars. Through a 2.4-inch, you'll see only half a dozen stars. Between two and three dozen stars show up through a 6-inch, with most of them concentrated in a hazy cloud at the cluster's eastern edge; there's also a nice chain of stars to the west. At 100x through a 10-inch scope, you'll count 75 or more stars.

Because you're already in the area, try finding **Hind's Variable Nebula** (NGC 1554/5). When the nebula shines at its brightest, observers have spotted it through 6-inch telescopes; at other times, it's invisible through a 16-inch scope. Also known as Struve's Lost Nebula, NGC 1554/5 glows because of radiation from the variable star T Tauri.

Lots of great observing awaits you in Monoceros. Messier listed one object here — **M50** — but easily could have included three more open clusters, all of which are brighter than M50. That's not to say M50 is weak. Even through a 2.4-inch telescope, you'll see two dozen stars, and, from a dark site, you might glimpse this magnitude 5.9 cluster with your naked eyes. At 100x through a 10-inch telescope, 150 stars pop into view within an area slightly smaller than the Full Moon. Note the yellow star on the cluster's southern edge and the void near the group's center.

The three open clusters brighter than M50 lie in a 15°-long line that runs north-south. Starting at the southern end, **NGC 2232** shines at

magnitude 3.9 not quite 2.5° north of Beta Monocerotis. Only about a dozen stars belong to this cluster, the brightest being 5th-magnitude 10 Monocerotis, which lies on the northern edge.

Nearly 12° north of NGC 2232 is **NGC 2244**. This magnitude 4.8 open cluster measures 20' across. Through a 6-inch telescope, you'll see 20 stars forming an oval elongated northwest to southeast. As you use larger scopes, progressively more stars will appear. It's difficult to tell, however, which belong to the cluster and which are background Milky Way stars.

Surrounding NGC 2244, but slightly offset, is the **Rosette Nebula** (NGC 2237–9). You'll get your best views of this object through low-power eyepieces. The Rosette measures 1° across, and, although faint, you can spot it through a 3-inch telescope. For a better view, however, use a 12-inch scope, wide-field eyepiece, and a nebula filter.

Now scan 5.5° north-northeast for the **Christmas Tree Cluster** (NGC 2264). Through a small telescope at 50x, a dozen or so stars extend to the east and west of 5th-magnitude 15 Monocerotis. This line forms the 0.5°-long base of the tree, whose top points southward. The southern stars of this asterism don't belong to the cluster. Larger telescopes will show a bright strip of nebulosity 5' long radiating westward from the brightest star. This is but one enhancement of the giant emission nebula Sharpless 2–273, which extends for 2° to the west. At the top of the Christmas Tree lies the Cone Nebula, an obscuring cloud of dust visible only through the biggest amateur telescopes.

Scan 1° south-southwest of the Christmas Tree Cluster, and you'll find the fascinating reflection nebula **NGC 2261**. Also known as Hubble's Variable Nebula, this object varies in brightness and structure over periods measured in days. NGC 2261 looks like no other object through a telescope. Through a 10-inch scope, a bright wedge 2' on a side radiates to the north of the variable star R Monocerotis. The nebula's cometary form

Designation	Right ascension	Declination	Magnitudes	Separation
BU 883	4h51m	11°03'	6.8, 7.0	15.1"
Iota Leporis	5h12m	−11°51'	4.5, 10.8	12.7"
Rho Orionis	5h13m	2°52'	4.6, 8.4	6.9"
Beta Orionis	5h15m	−8°12'	0.3, 10.4	9.5"
Lambda Orionis	5h35m	9°56'	3.7, 5.6	4.4"
Iota Orionis	5h35m	−5°55'	2.9, 7.0	11.4"
Sigma Orionis	5h39m	−2°35'	4.0, 7.5	12.9"
KUI 21	5h55m	11°46'	6.5, 12.0	22.7"
Epsilon Monocerotis	6h24m	4°35'	4.5, 6.5	12.9"
Beta Monocerotis	6h29m	−7°01'	4.7, 5.2, 6.1	7.2", 9.9"
STT 143	6h31m	16°55'	6.3, 9.4	8.0"
Nu¹ Canis Majoris	6h36m	−18°40'	5.8, 8.5	17.4"
30 Geminorum	6h44m	13°14'	4.6, 11.1	27.2"
Alpha Canis Majoris	6h45m	−16°43'	−1.5, 8.5	5.5"
38 Geminorum	6h55m	13°10'	4.8, 7.1	7.1"
BU 1060	6h59m	3°35'	6.0, 11.0	3.6"
Lambda Geminorum	7h18m	16°32'	3.6, 10.7	9.6"
Eta Canis Minoris	7h28m	6°56'	5.3, 11.1	4.0"
BUP 104	7h37m	−4°06'	5.1, 13.2	26.1"
STF 1143	7h48m	5°24'	7.0, 11.0	9.3"

▲ **REFLECTION NEBULA** M78 in Orion is the sky's brightest reflection nebula. It lies 2½° northeast of Alnitak (Zeta Orionis). CARSTEN FRENZL

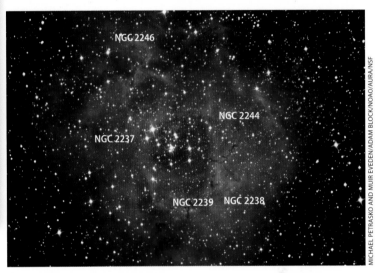

▲ **THE ROSETTE NEBULA** (NGC 2237–9) in Monoceros the Unicorn accounts for four entries in the NGC catalog, and this doesn't include the magnitude 4.8 star cluster at its center (NGC 2244).

appears uniformly bright with a sharply defined edge.

In Orion, there's really only one place to start: the **Orion Nebula** (M42). Visible to the naked eye as the middle "star" in Orion's "sword," this spectacular object looks great through any size telescope and at any magnification. Through a 2.4-inch scope, you'll see the Trapezium, a group of four stars (Theta¹ A, B, C, and D Orionis) that formed within the nebula's gas. If you increase your telescope aperture and magnification, you'll see up to five additional stars — Theta¹ E, F, G, and H, the last of which is a double star (see the map at right). When you view this region at low power, note the Fish's Mouth, an area of dark material that protrudes into the brightest part of M42.

Just north of the Fish's Mouth lies **M43**, which astronomers consider a separate object only for cataloging purposes. Look for the star NU Orionis in the center of this 15'-diameter object.

Move ½° north of the Orion Nebula to find the **Running Man Nebula** (NGC 1973/5/7). The two bright stars involved with the nebula are 42 Orionis (magnitude 4.6) and 45 Orionis (magnitude 5.2). Because the Running Man Nebula is a reflection nebula, observe it without a nebula filter. Its light is reflected starlight scattered throughout the gas and dust, not reddish light emitted by hydrogen (which a nebula filter transmits).

Point your telescope at Alnitak (Zeta Orionis). Move just 18' east-northeast to the **Flame Nebula** (NGC 2024). Normally, a large object like the Flame (30' across) would be easy to see, but the glare from magnitude 1.7 Alnitak interferes. Increase the magnification enough to capture the whole Flame and to place Alnitak out of the field of view to the west.

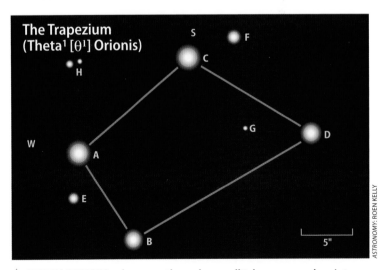

▲ **THETA¹ ORIONIS,** when seen through a small telescope, resolves into four stars called the Trapezium. Astronomers lettered the stars A, B, C, and D by right ascension, not brightness. Star A shines at magnitude 6.7, B at 8.0, C at 5.1, and D at 6.7. If your sky conditions are good, an 8-inch scope may reveal 11th-magnitude E and F. You probably will need a 14-inch telescope to find G and H, which both glow faintly at 15th magnitude.

MAP

11

6

↑ Map 5 ↑

8h

7h

GEMINI

ζ

λ

NGC 2395

NGC 2355

CANCER

Abell 21

STT 143

γ

J900

BL

ν

ξ

NGC 2194

NGC 2169

38

30

ξ

10°

β

NGC 2247

NGC 2264

NGC 2245

NGC 2141

μ

CANIS MINOR

γ   ε

β

NGC 2261

NGC 2251

η

13

Procyon

NGC 2236

STF 1143

α

NGC 2237

ε

NGC 2186

HYDRA

BU 1060

NGC 2244

Map 12

ζ

δ¹

18

0°

NGC 2324

NGC 2301

NGC 2346

δ

ζ

NGC 2286

M48

BUP 104

IC 466

NGC 2311

NGC 2232

NGC 2302

NGC 2182

MONOCEROS

γ

NGC 2316

β

NGC 2170

M50

-10°

NGC 2335

CRL 915

NGC 2506

NGC 2353

IC 2177

NGC 2525

NGC 2343

θ

NGC 2539

NGC 2374

W   Ced 90

Mel 71

NGC 2359

NGC 2345

μ

Min 1-18

NGC 2423

NGC 2360

γ

NGC 2438

M47

θ

M46

NGC 2414

α

CANIS MAJOR

Sirius

ι

β

PUPPIS

Sh2-301

ν³

ν¹

NGC 2204

ν²

-20°

8h

7h

↓ Map 17 ↓

EQUATORIAL REGION 2

This map displays constellations and deep-sky
objects during mid-winter north of the equator
and mid-summer in the Southern Hemisphere.

Constellation boundary	Constellation figure	Magnitudes	
		● Sirius	
		● 0.0	• 4.0
		● 1.0	• 5.0
		● 2.0	• 6.0
		● 3.0	• 7.0
			• 8.0

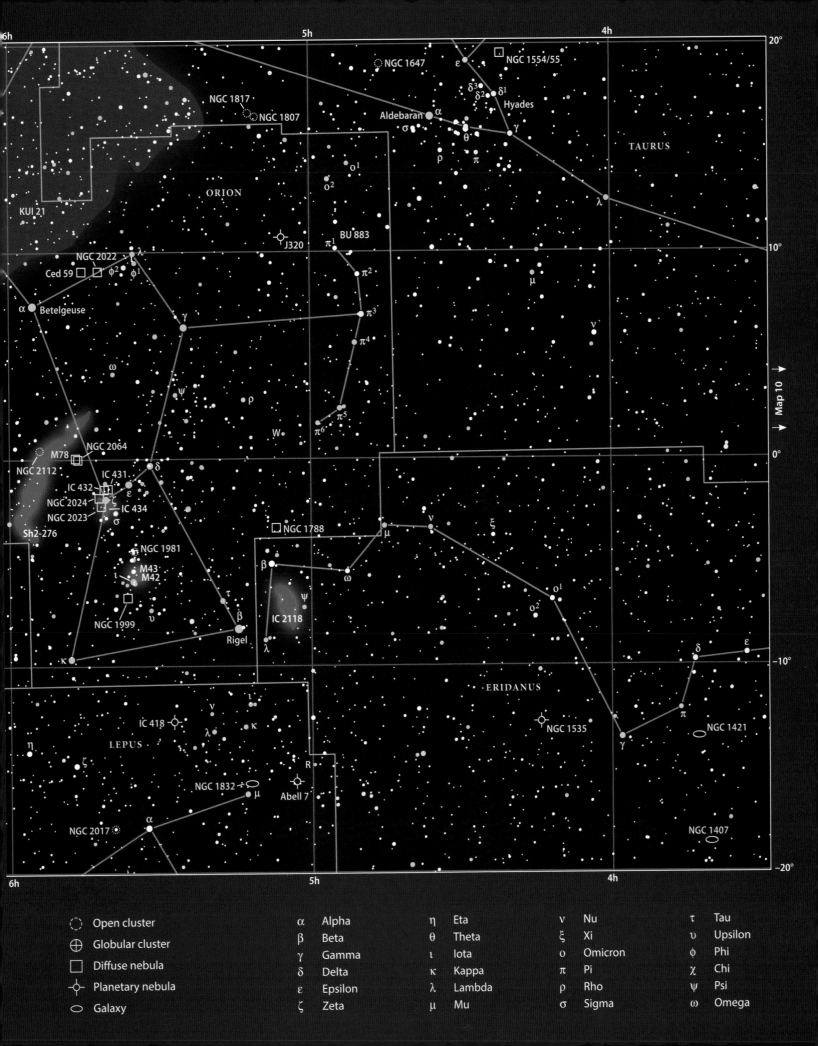

Map 10

Open cluster

Globular cluster

Diffuse nebula

Planetary nebula

Galaxy

α	Alpha	η	Eta	ν	Nu	τ	Tau
β	Beta	θ	Theta	ξ	Xi	υ	Upsilon
γ	Gamma	ι	Iota	ο	Omicron	φ	Phi
δ	Delta	κ	Kappa	π	Pi	χ	Chi
ε	Epsilon	λ	Lambda	ρ	Rho	ψ	Psi
ζ	Zeta	μ	Mu	σ	Sigma	ω	Omega

MAP **12** **EQUATORIAL REGION 3**

TOM BASH AND JOHN FOX/ADAM BLOCK/NOAO/AURA/NSF

▲ **THE BEEHIVE CLUSTER (M44) in Cancer makes a fine binocular target. From a dark site, you'll see it easily with naked eyes.**

The Lion's galaxies

Leo the Lion dominates the next map. No fewer than five Messier objects — all bright galaxies — await you there.

Before you begin hunting galaxies, however, pull out your binoculars and look toward Cancer the Crab at the open cluster called the **Beehive** (M44). Binoculars labeled 7x50 will show you the Beehive's overall structure, but 15x or higher binoculars reveal star patterns that resemble arcs or chains. For best results, mount your binoculars on a sturdy camera tripod. If you choose to observe M44 through a telescope, use low power.

Next, focus on **M67** in Cancer. Through binoculars, you might mistake M67 for a magnitude 6.9 globular cluster. It is, in fact, an open cluster that large telescopes show to be 30' across — the same size as the Full Moon. Through an 8-inch scope, you'll see 50 stars, and nearly double that through a 12-inch instrument.

In Leo, light from brilliant Regulus (Alpha Leonis) hides one of the Milky Way's dwarf companion galaxies, **Leo I** (it doesn't have an NGC number). This object shines at magnitude 10.2 — bright for a galaxy, but more than 4,300 times fainter than Regulus. Leo I lies 20' due north of Regulus, so place the bright star out of your field of view to the south, and

look for a moderate brightening of the background roughly 8' across.

While you're in Leo, scan 1.5° south of Lambda Leonis to find **NGC 2903** (Map 6). Observers often overlook this spiral galaxy in favor of Leo's five Messier objects. NGC 2903, however, is brighter than all but M65. A 10-inch scope shows a halo measuring 4' by 2' around a bright core.

Midway between Gamma and Zeta Leonis lies Hickson 44, a group of four galaxies centered on **NGC 3190** (Map 6). Hickson 44 is one of 100 compact galaxy groups in a catalog compiled in the 1980s by Canadian astronomer Paul Hickson. Magnitude 11.2 NGC 3190 is the brightest member of the group, which also includes NGC 3185, NGC 3187, and NGC 3193. You'll need at least a 10-inch telescope to see all four.

Less than 1° east of magnitude 2.0 Gamma Leonis lies magnitude 10.3 **NGC 3227**. The galaxy is a fine target on its own, but what makes it special is its interaction with NGC 3226, a magnitude 12.3 elliptical galaxy seemingly attached to NGC 3227's northern end. NGC 3227 appears nearly round, measuring 4.1' by 3.9'.

Draw a line between Rho and Theta Leonis, then travel 4° along this line to three bright Messier objects. These galaxies — **M95**, **M96**, and **M105** — and six others belong to the M96 Group. You can view M95, M96, and M105 together if you use an eyepiece that gives at least a 1.5°

▲ SPIRAL GALAXY M65 in Leo shines at magnitude 8.8 and measures 10' by 2.7'. It lies 35 million light-years away.

◄ SPIRAL GALAXY NGC 3521 in Leo can be spotted through binoculars, but to see the most detail, use a telescope at medium or high power.

▲ BARRED SPIRAL GALAXY M95 in Leo is a member of the Leo I group, which also contains M96 and M105. This group lies approximately 38 million light-years away. M95 measures 4.4' by 3.3'.

/// **DOUBLE-STAR DELIGHTS — MAP 12**

Designation	Right ascension	Declination	Magnitudes	Separation
HO 350	8h04m	12°11'	6.7, 10.8	4.8"
Epsilon Hydrae	8h47m	6°25'	3.5, 6.8	3.2"
Rho Hydrae	8h48m	5°49'	4.4, 11.9	12.4"
Struve 1295	8h55m	−7°57'	6.7, 6.9	4.2"
Alpha Cancri	8h59m	11°51'	4.3, 11.8	10.9"
Theta Hydrae	9h14m	2°19'	3.8, 9.8	27.1"
Gamma Leonis	10h20m	19°50'	2.6, 3.8	4.5"
Chi Leonis	11h05m	7°20'	4.7, 11.0	3.6"
Gamma Crateris	11h25m	−17°41'	4.1, 9.6	5.3"
83 Leonis	11h27m	3°02'	6.5, 7.6	40.8"
90 Leonis	11h35m	16°47'	6.1, 7.4	3.4"

field of view. M95 is a spiral galaxy that shines at magnitude 9.7. M96 is also a spiral, but it's a half-magnitude brighter at 9.2. Just a tad fainter, elliptical galaxy M105 glows at magnitude 9.3.

Three more bright galaxies — the Leo Triplet — lie 2.5° southeast of Theta Leonis, midway between it and Iota Leonis. An eyepiece with a 1° field of view encompasses all three, but you'll want to crank up the power on each to examine its details.

First is spiral galaxy **M65**, which marks the Triplet's southwestern corner. M65 shines at magnitude 8.8 and measures four times as long as it is wide. Through a 10-inch or larger telescope, look for irregular structure near M65's core. This galaxy appears somewhat inclined to the line of sight. Measurements show M65 tilts 15° from being classified as edge-on.

At magnitude 9.0, spiral galaxy **M66** ranks as one of the 20 brightest galaxies in the sky. M66 measures 8' by 4', and its arms wrap tightly around its core. Use at least a 12-inch telescope to pick out its arms.

NGC 3628 completes the triplet and appears much fainter than M65 or M66. At magnitude 9.5, it's really not that much dimmer than its neighbors, but its light spreads over an area measuring 14' by 4'. Look for a faint dust lane south of NGC 3628's center.

▲ M66 IN LEO is a magnitude 9.0 spiral galaxy, which measures 9' by 4'. Along with M65 and NGC 3628, it forms the Leo Triplet.

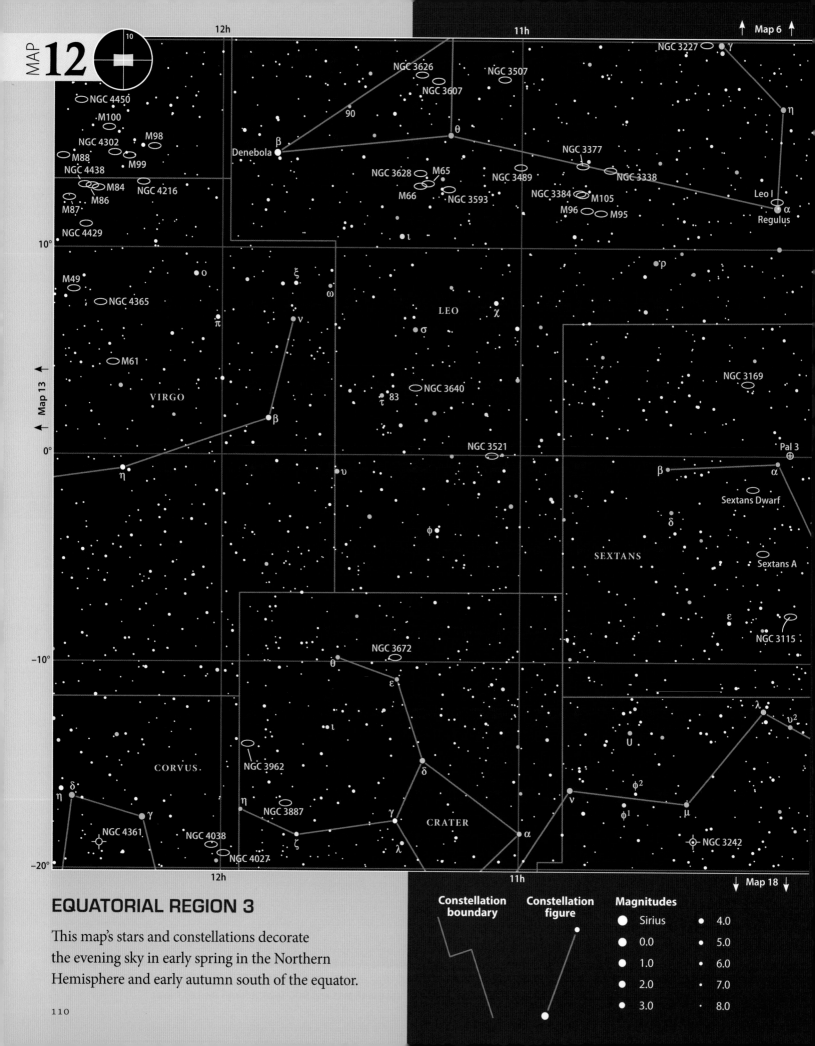

MAP 12

10

12h

11h

NGC 3227

γ

NGC 3626

NGC 3507

η

NGC 3607

90

θ

NGC 3377

β

NGC 3489

Denebola

NGC 3628

M65

NGC 3338

NGC 3593

Leo I

M66

NGC 3384

M105

α

NGC 3596

M96

M95

Regulus

ι

10°

ρ

M49

o

ξ

M13

ω

NGC 4365

π

LEO

χ

ν

σ

VIRGO

M61

83

NGC 3640

τ

β

NGC 3169

Pal 3

η

0°

NGC 3521

β

α

υ

Sextans Dwarf

δ

φ

SEXTANS

Sextans A

ε

NGC 3115

−10°

NGC 3672

θ

ε

λ

υ²

U

ι

φ²

NGC 3962

δ

CORVUS.

δ

φ¹

μ

η

γ

ν

η

NGC 3887

γ

φ¹

δ

γ

CRATER

−20°

NGC 4361

ζ

α

NGC 3242

NGC 4038

λ

NGC 4027

12h

11h

EQUATORIAL REGION 3

This map's stars and constellations decorate
the evening sky in early spring in the Northern
Hemisphere and early autumn south of the equator.

Constellation boundary	Constellation figure	Magnitudes	
		Sirius	4.0
		0.0	5.0
		1.0	6.0
		2.0	7.0
		3.0	8.0

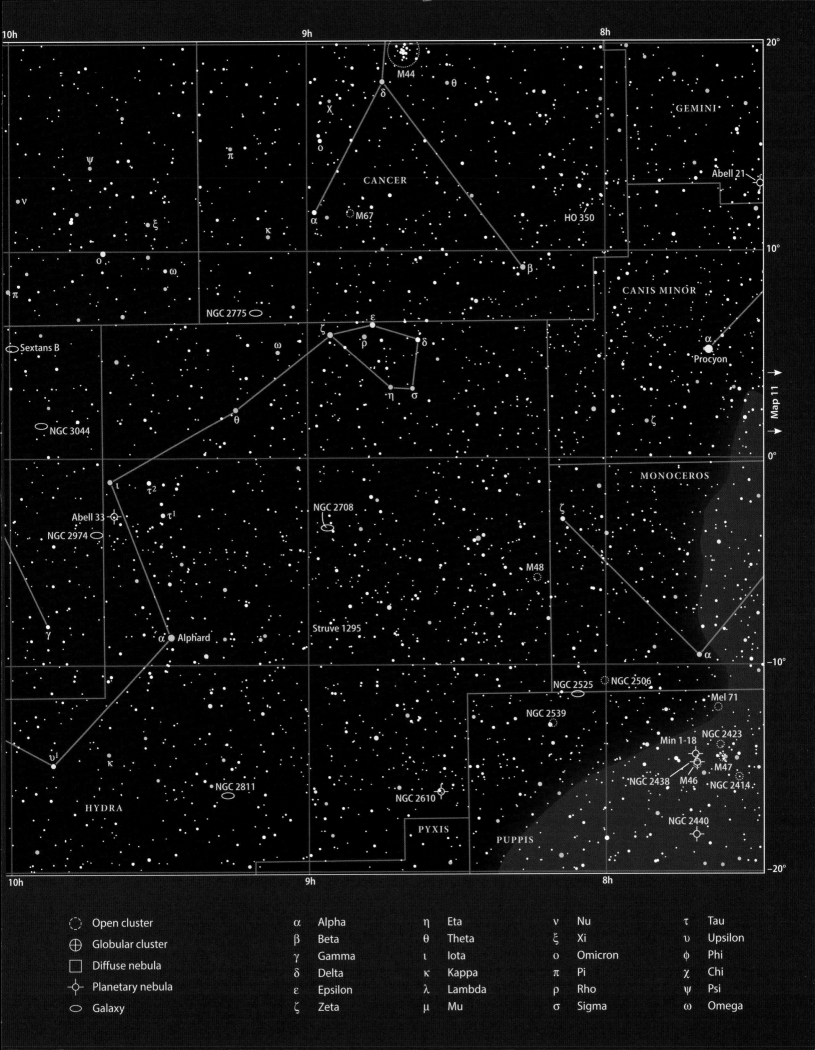

20°

GEMINI

M44

θ

δ

χ

CANCER

ο

Abell 21

α M67

HO 350

β

10°

CANIS MINOR

ζ ε

ρ δ

ω

α

Procyon

NGC 2775

η σ

Sextans B

θ

ζ

NGC 3044

0°

Map 11

MONOCEROS

ι

τ²

NGC 2708

ζ

Abell 33

τ¹

NGC 2974

M48

α

Struve 1295

γ

α Alphard

–10°

NGC 2525 NGC 2506

Mel 71

NGC 2539

υ¹

NGC 2423

Min 1-18

κ

M47

NGC 2811

NGC 2438 M46 NGC 2414

NGC 2610

NGC 2440

HYDRA

PYXIS

PUPPIS

–20°

⬭	Open cluster	α	Alpha	η	Eta	ν	Nu	τ	Tau
⊕	Globular cluster	β	Beta	θ	Theta	ξ	Xi	υ	Upsilon
☐	Diffuse nebula	γ	Gamma	ι	Iota	ο	Omicron	φ	Phi
✧	Planetary nebula	δ	Delta	κ	Kappa	π	Pi	χ	Chi
⬭	Galaxy	ε	Epsilon	λ	Lambda	ρ	Rho	ψ	Psi
		ζ	Zeta	μ	Mu	σ	Sigma	ω	Omega

MAP **13** EQUATORIAL REGION 4

◀ **GLOBULAR CLUSTER M5** in Serpens appears as a fuzzy magnitude 5.7 star to the unaided eye. Through a telescope, it explodes with detail.

▼ **SPIRAL GALAXY M100** in Coma Berenices lies in a rich star field. Through a 12-inch telescope, you'll see hundreds of galaxies nearby.

Realm of the nebulae

Nineteenth-century observers called the area covered by Map 13 the "realm of the nebulae." They weren't describing nebulae in the current sense, however. Their nebulae were galaxies, which looked nebulous through the small telescopes most observers used. Luckily, amateur scopes today are bigger, better, and easier to acquire. Unfortunately, our sky is brighter due to light pollution.

Roughly 9° east-southeast of Epsilon Virginis, you'll find **NGC 5248**. This magnitude 10.3 spiral galaxy actually resides in the southwest corner of Boötes and is that constellation's brightest galaxy. NGC 5248's bright core and outer regions show up even through small telescopes. A 12-inch scope shows a barely elongated oval 3.5' by 3' with a stellar nucleus slightly offset to the north. You'll also see a prominent dark patch south of the core and some of the structure making up the spiral arms. If you're observing with an even larger scope, look for two of NGC 5248's 15th-magnitude companion galaxies: UGC 8575 lies 27' to the west, and UGC 8629 is 30' to the southeast.

Of the 88 constellations, Serpens the Serpent is the only one that's not contiguous. Stellar cartographers call its western part, or head, Serpens Caput and its eastern part, or tail, Serpens Cauda. Map 13 contains most of Serpens Caput. Find Serpens Cauda on Map 14.

The brightest deep-sky object in Serpens Caput is globular cluster **M5**. You'll see M5 without optical aid from a dark site — it shines at magnitude 5.7. The almost equally bright star 5 Serpentis lies 22' to the southeast; M5 looks fuzzier. Through a telescope, however, M5 really

stands out. Through a 6-inch scope, the cluster appears 10' in diameter with a densely packed center about one-quarter its diameter. Through a 10-inch telescope, individual stars appear to form streamers that cross the cluster and radiate from it.

The northern part of Libra is bereft of deep-sky objects within the reach of small scopes. If your telescope measures 12 inches or larger, however, many galaxies lie beyond the magnitude 11 threshold. Among them are **NGC 5728**, **NGC 5812**, and **NGC 5878**.

Approximately 2° southeast of Eta Crateris, you'll find the **Antennae** (NGC 4038 and NGC 4039). The bright tails of these interacting galaxies are huge plumes of material thrown out by tidal interactions. This pair is visible, although indistinct, through a 6-inch telescope. Move up to a 12-inch scope, and you'll notice NGC 4038 is the brighter of the two. Both cores will be visible at 100x through a 12-inch scope. If you increase the magnification to 200x, you'll see bright knots, mainly in NGC 4038. If you want to see the tails, attend a spring star party and hope someone there has a 24-inch telescope.

Below and nearly equidistant from Delta and Gamma Corvi lies planetary nebula **NGC 4361**. This magnitude 10.9 object shows up well even through a 6-inch scope because its light concentrates in an area only about 1' across. Increase the magnification, and you'll see NGC 4361's irregular edge and relatively bright central star.

Moving to Coma Berenices, it's hard to believe it holds so many deep-sky treasures. Map 13 contains only the southern half of the constellation, but even that small area holds seven Messier objects.

Two globular clusters populate Coma's southeastern corner. **M53** lies 1°

northeast of Alpha Comae Berenices, and you'll find **NGC 5053** 1.5° east of the star. M53 shines at magnitude 7.7 and measures about 12' across. A 6-inch telescope will resolve its outer stars well and show the core as broad and dense. In contrast, magnitude 9.9 NGC 5053 is one of the least concentrated globulars, looking somewhat like a tight open cluster. Even a 12-inch scope shows only about 30 stars in a 5'-wide area.

Just 2° east-southeast of 11 Comae lies spiral galaxy **NGC 4450**. Through a 10-inch scope, this magnitude 10.1 object covers roughly 4' by 3'. You'll see a 9th-magnitude star 4' to the southwest. NGC 4450's core appears lumpy with a starlike nucleus slightly off-center to the east.

Finally, we come to Virgo. If you're a galaxy-hunter, you can spend a whole season in this one constellation. Virgo holds more bright galaxies than any other constellation.

Start about 4.5° south-southeast of Gamma Virginis at elliptical galaxy **NGC 4697**. At magnitude 9.3, this is a bright object, but detail is lacking because of its distance. Through a 10-inch telescope, NGC 4697 appears 2' by 1' with a large, bright core and a faint halo.

Move 3° south from NGC 4697 to find **NGC 4699**. At first glance through a 6-inch scope, this object looks like an elliptical galaxy similar to NGC 4697. In fact, it's a tightly wound spiral galaxy. Use a 12-inch or larger telescope and high power to see its arms. NGC 4699 measures 3.5' by 2.5' and glows brightly (for a galaxy) at magnitude 9.6.

Only one non-galaxy, non-stellar deep-sky object brighter than magnitude 12 resides in Virgo, and it's a worthwhile target — globular cluster **NGC 5634**. To find it, point your scope midway between Mu and Iota Virginis. In addition to the cluster, a magnitude 8.5 foreground star lurks only 1.3' east-southeast of NGC 5634's center. NGC 5634 shines at magnitude 9.4 and measures about 5' across.

/// **DOUBLE-STAR DELIGHTS — MAP 13**

Designation	Right ascension	Declination	Magnitudes	Separation
Beta Virginis	11h51m	1°45'	3.8, 8.8	12.3"
Delta Corvi	12h30m	−16°31'	3.1, 9.3	24.1"
Gamma Virginis	12h42m	−1°26'	3.6, 3.7	3.7"
Theta Virginis	13h10m	−5°32'	4.4, 9.4	7.1"
54 Virginis	13h13m	−18°49'	6.8, 7.3	5.3"
STF 1750	13h30m	−6°26'	6.1, 11.4	29.8"
Phi Virginis	14h28m	−2°13'	5.0, 9.5	5.1"
BU 1085	14h59m	−4°59'	6.1, 13.2	9.4"
18 Librae	14h59m	−11°08'	6.0, 10.2	19.8"
5 Serpentis	15h19m	1°47'	5.2, 10.2	11.3"
6 Serpentis	15h21m	0°43'	5.5, 10.1	3.1"
Delta Serpentis	15h35m	10°32'	4.2, 5.2	4.0"

ADAM BLOCK/NOAO/AURA/NSF

/// **MESSIER OBJECTS IN COMA BERENICES AND VIRGO**

Object	Type	Magnitude
M49	Elliptical galaxy	8.4
M53	Globular cluster	7.7
M58	Spiral galaxy	9.6
M59	Elliptical galaxy	9.6
M60	Elliptical galaxy	8.8
M61	Spiral galaxy	9.6
M64	Spiral galaxy	8.5
M84	Elliptical galaxy	9.1
M85	Spiral galaxy	9.1
M86	Elliptical galaxy	8.9
M87	Elliptical galaxy	8.6
M88	Spiral galaxy	9.6
M89	Elliptical galaxy	9.7
M90	Spiral galaxy	9.5
M91	Spiral galaxy	10.1
M98	Spiral galaxy	10.1
M99	Spiral galaxy	9.9
M100	Spiral galaxy	9.3
M104	Spiral galaxy	8.0

PAUL AND DANIEL KOBLAS/ADAM BLOCK/NOAO/AURA/NSF

▲ **M90 IN VIRGO** spans 9.5' by 4.5' and offers impressive detail through an 8-inch telescope. Through a larger scope, fainter background galaxies appear.

▲ **THE EYES** (NGC 4438 [left] and NGC 4435) lie in the Virgo cluster. These interacting galaxies glow at magnitudes 10.2 and 10.8, respectively.

◄ **SPIRAL GALAXY** M61 in Virgo is similar in size to our Milky Way. This magnitude 9.6 giant spans 100,000 light-years. ADAM BLOCK/NOAO/AURA/NSF

◄ **THE SOMBRERO GALAXY** (M104) is a favorite target among observers. Look for the dust and cold gas that split this object lengthwise. CARSTEN FRENZL

MAP
13
14

Abell 2151

16h

κ

NGC 5962

γ
β
φ
υ

χ

ω

IC 4593

δ

15h

ξ

ι

o
π

ζ

BOÖTES

↑ Map 7 ↑

α
Arcturus

10°

λ

α

SERPENS CAPUT

HERCULES

ε

NGC 5921

ψ

ω

M5
5

NGC 5846

109

τ

Map 14 ←

←

λ

σ

6

Pal 5

NGC 5792

φ
υ

0°

NGC 6118

δ

ε

μ

BU 1085

μ

NGC 5634

OPHIUCHUS

υ

NGC 5812

LIBRA

δ

β

ι

-10°

ψ

χ

ξ

ε

NGC 5885

18
ξ²
ξ¹

κ

λ

μ

NGC 5878

SCORPIUS

γ

η

θ

o

ζ

ν

α

IC 972

φ

NGC 5728

χ

ν
β

κ

ι

16h

15h

↓ Map 19 ↓

-20°

EQUATORIAL REGION 4

View the constellations and deep-sky objects on
this map during early summer in the Northern
Hemisphere and early winter south of the equator.

Constellation boundary	Constellation figure	Magnitudes	
		Sirius	4.0
		0.0	5.0
		1.0	6.0
		2.0	7.0
		3.0	8.0

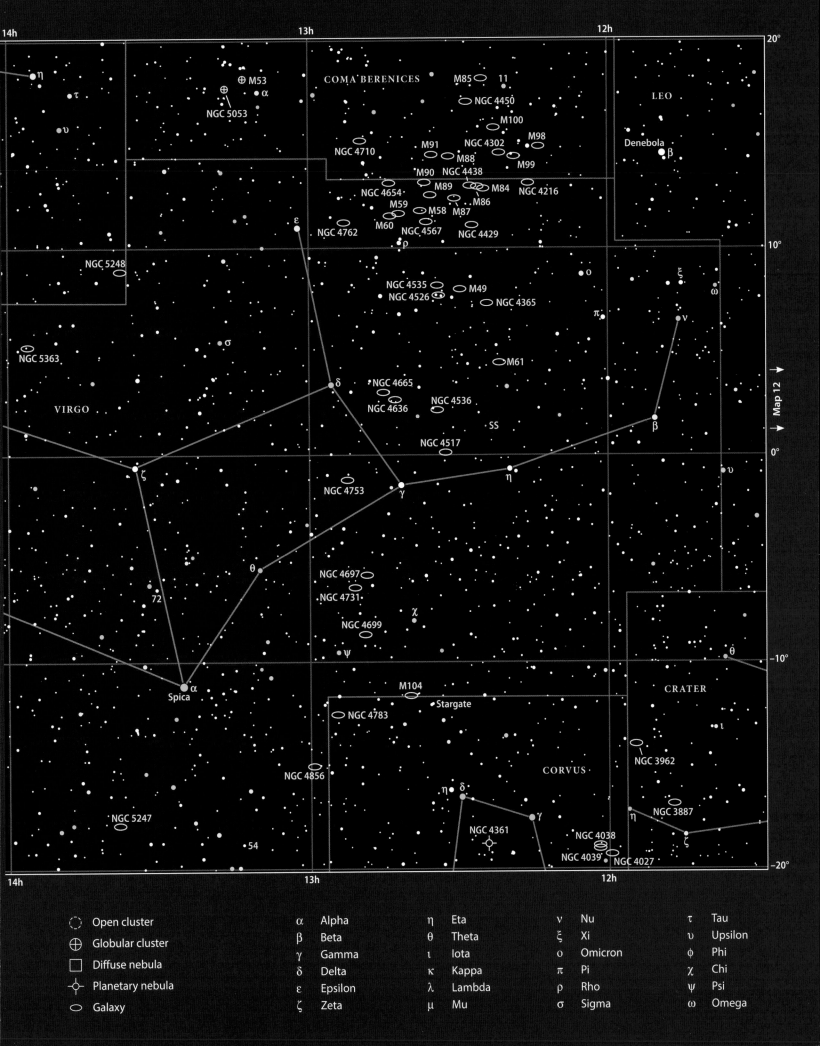

Open cluster
Globular cluster
Diffuse nebula
Planetary nebula
Galaxy

α	Alpha	η	Eta	ν	Nu	τ	Tau
β	Beta	θ	Theta	ξ	Xi	υ	Upsilon
γ	Gamma	ι	Iota	ο	Omicron	φ	Phi
δ	Delta	κ	Kappa	π	Pi	χ	Chi
ε	Epsilon	λ	Lambda	ρ	Rho	ψ	Psi
ζ	Zeta	μ	Mu	σ	Sigma	ω	Omega

MAP 14   EQUATORIAL REGION 5

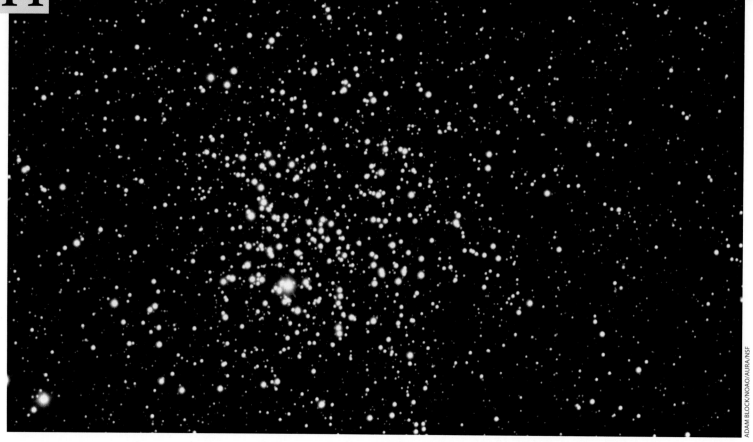

ADAM BLOCK/NOAO/AURA/NSF

▲ **THE WILD DUCK CLUSTER** (M11) swarms with stars. Use 80x to 100x through an 8-inch or larger telescope for the best view.

Globular central

If Map 13 contains the realm of the galaxies, then the next map encompasses the realm of globular clusters. Our Milky Way contains roughly 200 globular clusters, and a third of them can be found in just three of the 88 constellations: Scorpius, Sagittarius, and Ophiuchus. (Ophiuchus occupies a large part of the next map. Find Scorpius and Sagittarius on Map 20.) But this region contains more than just globulars.

Because the Milky Way passes through Sagitta, Aquila, Scutum, Serpens, and part of Ophiuchus, open clusters and nebulae abound here. You won't find many galaxies, however. The reason is the composition of the Milky Way itself. In addition to stars, immense diffuse gas clouds and dust permeate our galaxy's spiral arms (which we see at night as the Milky Way). This material blocks out light from more distant stars and galaxies. The drop-off in numbers of observable galaxies is so extreme that early 20th-century astronomers referred to this area as the Zone of Avoidance.

In Sagitta, find globular cluster **M71** midway between Gamma and Delta Sagittae. This 8th-magnitude cluster looks irregular at low power because of foreground stars. Through a 6-inch scope, you can resolve a couple dozen stars, and the view doesn't improve much through a 12-inch — only 50 stars are visible within a 4' area.

Below Sagitta, Aquila provides an opportunity to view an easily recognized dark nebula. About 1° west of Gamma Aquilae lies a complex of dark nebulae known as Barnard's E. Dark nebulae are not voids, as astronomers once believed, but rather clouds of dust and cold gas that block out light from more distant stars. American astronomer Edward Emerson Barnard cataloged hundreds of these objects strewn across the bright background of the Milky Way.

Use binoculars or a low-power eyepiece in your telescope to see Barnard 143, which forms a prominent C shape roughly 20' long oriented east-west. Just to the south lies Barnard 142, a ½°-long stretch of dark nebulosity. Together, these two form the E, which stands out well because the arms of our galaxy that form the background are full of faint stars.

Not quite 4° north-northwest of Delta Aquilae lies planetary nebula **NGC 6781**. Through an 8-inch telescope, this magnitude 11.4 object spans more than 1.5' and stands out well against a star-filled background. For best results, use a 6-inch or larger telescope and a nebula filter, and look for NGC 6781's irregularly illuminated outer edge.

Now look a bit more than 4° southwest of Delta Aquilae for globular cluster **NGC 6760**. Glowing at magnitude 9.1, NGC 6760 is a concentrated globular cluster you easily will spot through a 3-inch scope. Even a 12-inch telescope doesn't resolve more than a handful of the globular's stars, but this size scope does show the cluster as mottled.

Southwest of Aquila lies the small constellation Scutum the Shield. Aquila contains six times more area than Scutum, but Aquila contains no Messier objects, and Scutum has two.

Nearly 2° southeast of Beta Scuti lies the **Wild Duck Cluster** (M11). This magnificent open cluster measures about 12' across and shines at magnitude 5.8, making it visible to your naked eyes from a dark site. M11 acquired its odd name when 19th-century observer William Henry Smyth wrote that it resembled a flock of wild ducks in flight. Smyth was referring to the cluster's triangular shape. Through a 6-inch telescope, you'll see more than 100 stars, many of them packed into the dense core. The brightest star shines at 8th magnitude and lies near M11's center.

Designation	Right ascension	Declination	Magnitudes	Separation
KUI 70	16h06m	−6°09'	6.5, 12.3	9.3"
Kappa Herculis	16h08m	17°03'	6.3, 6.5	28.1"
Omega Herculis	16h25m	14°02'	4.5, 11.0	28.4"
37 Herculis	16h41m	4°13'	5.8, 7.0	9.7"
19 Ophiuchi	16h47m	2°04'	6.1, 9.4	23.4"
Alpha Herculis	17h15m	14°23'	3.5, 5.4	4.8"
61 Ophiuchi	17h45m	2°35'	6.2, 6.6	20.6"
Struve 2325	18h31m	−10°48'	5.8, 9.3	2.3"
Delta Scuti	18h42m	−9°03'	4.7, 12.2	15.2"
5 Aquilae	18h46m	−0°58'	5.7, 6.1, 7.9	12.8", 26.3"
11 Aquilae	18h59m	13°37'	5.4, 8.9	17.8"
HO 275	19h51m	−10°46'	5.4, 13.6	21.1"
Beta Aquilae	19h55m	6°24'	3.9, 11.8	12.9"
Struve 2644	20h13m	0°52'	6.9, 7.1	2.7"

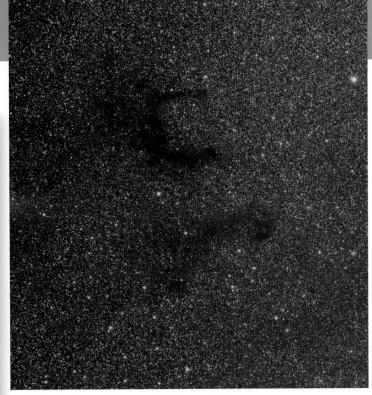

▲ BARNARD'S E (B142 and B143) lies a bit more than 3° northwest of Altair (Alpha Aquilae). Use binoculars or a wide-field eyepiece in your telescope to view this area. KEES SCHERER

▲ PLANETARY NEBULA NGC 6572 in Ophiuchus shines at magnitude 8.1. You won't see much detail, but NGC 6572's green color makes it a popular target. BRUCE BODNER/ADAM BLOCK/NOAO/AURA/NSF

▲ M12 in Ophiuchus is one of the least concentrated globular clusters. Use high magnification to compare it to M10, at right. MICHAEL GARIEPY/ADAM BLOCK/AURA/NSF

▲ GLOBULAR CLUSTER M10 in Ophiuchus shines at magnitude 6.6, making it one of the sky's brightest globulars. MICHAEL AND MICHAEL McGUIGGAN/ADAM BLOCK/NOAO/AURA/NSF

M26 is Scutum's other Messier object. M26 is also an open cluster, but fainter (magnitude 8.0) and looser than M11. A 4-inch telescope reveals 25 stars within a 10' area, and a 12-inch scope will triple that number.

The lone galaxy of note on Map 14 — **Barnard's Galaxy** (NGC 6822) — is also a challenge to see. The galaxy's magnitude of 8.8 is misleading because the light spreads out over an area 19' by 15'. Not many bright stars reside near NGC 6822. This object lies more than 6° northeast of Rho¹ Sagittarii. Use low power under a dark sky, and look for a roughly rectangular haze slightly brighter than the background.

Ophiuchus contains no fewer than 10 globular clusters brighter than magnitude 9. Seven belong to Messier's catalog: **M9**, **M10**, **M12**, **M14**, **M19** (Map 20), **M62** (Map 20), and **M107**. Three do not: magnitude 8.2 **NGC 6293** (Map 20), magnitude 8.2 **NGC 6356**, and magnitude 8.9 **NGC 6366**. Each of these globular clusters offers a unique observing experience. For example, compare M10 and M12. Move from one to the other and back again. View them at low power against a wide background. Then, view them at as high of a magnification as conditions allow. By noting the similarities and differences between the two views, you'll become a better observer.

▲ THE EAGLE NEBULA (M16) is a bright emission nebula associated with a prominent open star cluster (not shown). Use an OIII filter for best results. BILL LOFQUIST/ADAM BLOCK/NOAO/AURA/NSF

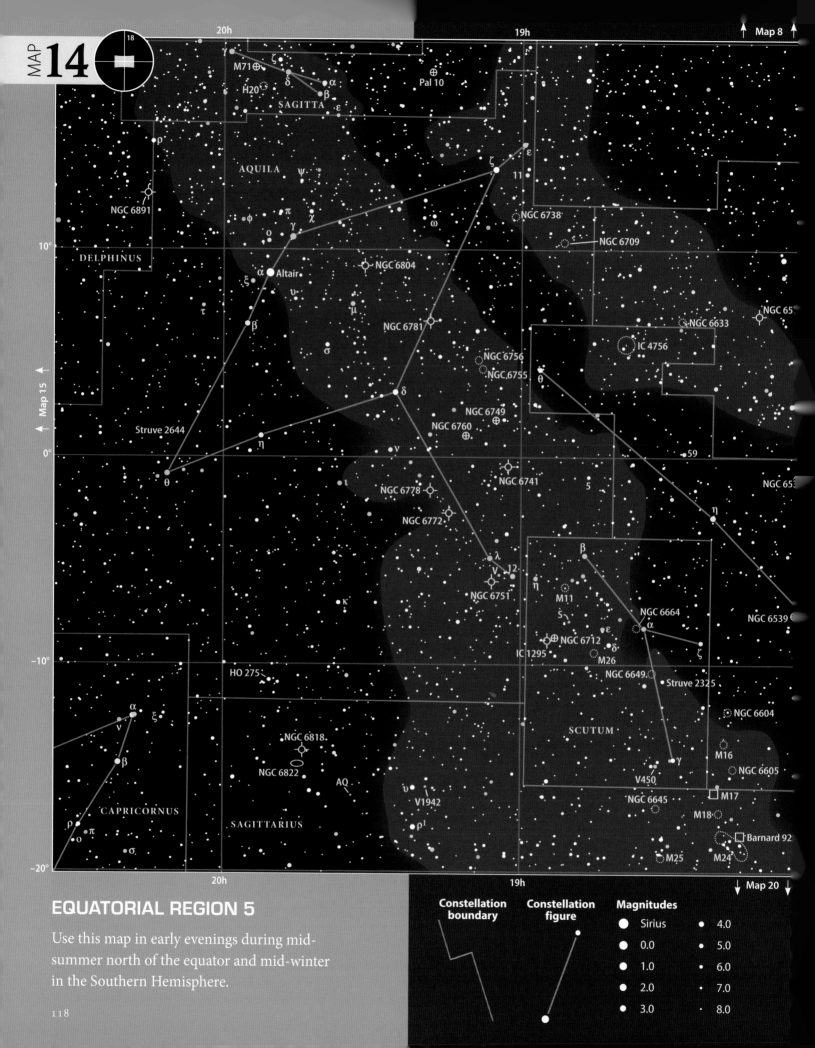

MAP
14
18
↑ Map 8 ↑

20h

19h

γ
ζ
M71⊕
δ
α
H20
β
ε
SAGITTA

Pal 10

ρ

AQUILA
ψ

NGC 6891

φ
π
γ
χ

ζ

ε
11

NGC 6738

NGC 6709

10°

o

α Altair
ξ
υ

ω

NGC 6804

NGC 65

DELPHINUS

τ
β

μ

NGC 6781

NGC 6633
IC 4756

NGC 6756
NGC 6755

θ

Map 15

Struve 2644
η

σ

ι

NGC 6749
NGC 6760

NGC 6741

5

59

NGC 65

0°

θ

ν

NGC 6778

NGC 6772

λ

ν 12
NGC 6751

β

η

M11
ζ

NGC 6664
α

η

NGC 6539

-10°

HO 275

NGC 6712
IC 1295
ε
δ
M26

NGC 6649
Struve 2325

ζ

NGC 6604

α
ν
ξ

NGC 6818

SCUTUM

M16
NGC 6605

β

NGC 6822

AQ

γ
V450

M17

CAPRICORNUS

υ
V1942

NGC 6645

M18

ρ
o π

SAGITTARIUS

ρ¹

Barnard 92

σ

M25
M24

-20°

20h

19h

↓ Map 20 ↓

EQUATORIAL REGION 5

Use this map in early evenings during mid-
summer north of the equator and mid-winter
in the Southern Hemisphere.

Constellation boundary	Constellation figure	Magnitudes	
		● Sirius	● 4.0
		● 0.0	· 5.0
		● 1.0	· 6.0
		● 2.0	· 7.0
		● 3.0	· 8.0

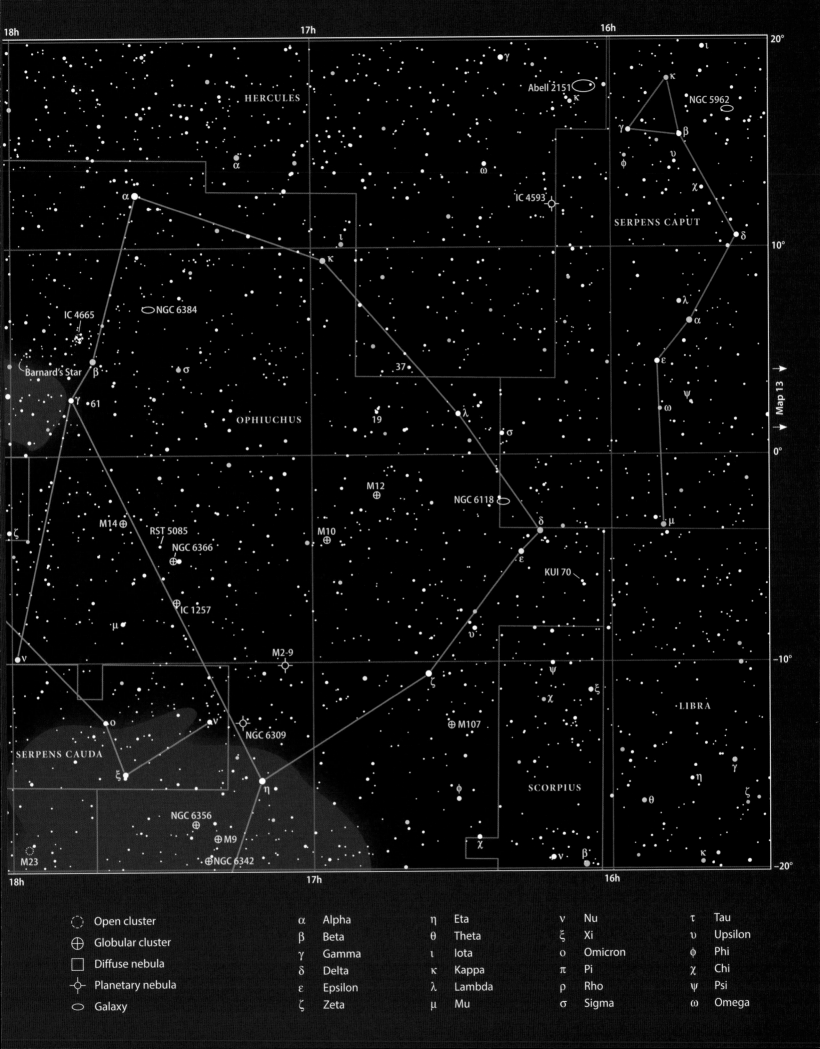

HERCULES

γ

Abell 2151

κ

NGC 5962

ι

κ

α

ω

φ

γ

β

υ

IC 4593

χ

SERPENS CAPUT

δ

10°

α

IC 4665

NGC 6384

κ

λ

Barnard's Star

β

α

37

ε

γ

61

ψ

OPHIUCHUS

ω

19

λ

μ

σ

ζ

M12

σ

0°

M14

NGC 6118

M10

δ

RST 5085

ε

NGC 6366

KUI 70

IC 1257

μ

υ

ψ

-10°

ν

M2-9

ξ

χ

LIBRA

o

ν

NGC 6309

M107

γ

SERPENS CAUDA

ξ

η

φ

η

SCORPIUS

θ

ζ

NGC 6356

M9

χ

κ

NGC 6342

ν

β

M23

Map 13

⬭	Open cluster	α	Alpha	η	Eta	ν	Nu	τ	Tau
⊕	Globular cluster	β	Beta	θ	Theta	ξ	Xi	υ	Upsilon
□	Diffuse nebula	γ	Gamma	ι	Iota	o	Omicron	φ	Phi
⊙	Planetary nebula	δ	Delta	κ	Kappa	π	Pi	χ	Chi
⬯	Galaxy	ε	Epsilon	λ	Lambda	ρ	Rho	ψ	Psi
		ζ	Zeta	μ	Mu	σ	Sigma	ω	Omega

MAP 15 EQUATORIAL REGION 6

PAT AND CHRIS LEE/ADAM BLOCK/NOAO/AURA/NSF

▲ **M2 IN AQUARIUS** is one of the sky's richest and most compact globular clusters. Through a telescope, M2 appears slightly elliptical.

Swimming with stars

The next star map includes part of the sky's so-called watery region, so called by early observers. It contains Delphinus the Dolphin and the northern parts of Aquarius the Water-bearer, Capricornus the Sea Goat, Cetus the Whale, and some of Pisces the Fish. Not shown — but nearby — are Piscis Austrinus the Southern Fish and Eridanus the River. Combined, these constellations cover more than 10 percent of the sky.

In Delphinus, the standout deep-sky object is globular cluster **NGC 6934**. This object shines at magnitude 8.7. Find it by dropping 4° south from Epsilon Delphini. NGC 6934 measures more than 8' across, but to see it to that extent, you'll need a 16-inch telescope. More modest instruments show it as 3' in diameter. You may pick out a few outlying stars, but its central region remains unresolved.

Apart from a few double stars, Equuleus the Little Horse contains no deep-sky objects of interest. But the other horse — Pegasus — boasts quite a few. The best is globular cluster **M15**. This magnitude 6.3 object measures more than 10' across. You can spot M15 with your naked eyes, but don't confuse it with the star SAO 107195, which lies only 17' to the east. Through a 6-inch or larger telescope, you'll see several hundred stars in a variety of patterns scattered about M15's dense, unresolved core.

Aquarius boasts three Messier objects. **M72** lies nearly 3.5° south-southeast of magnitude 3.8 Epsilon Aquarii. M72 glows at magnitude 9.2 and measures approximately 5' across. The cluster is so dense that

you won't be able to resolve any stars near its core.

Only 1.3° east of M72 lies **M73**. Most lists classify M73 as an open cluster, but it consists of only four stars: a nearly equilateral triangle of 10th- and 11th-magnitude stars with a fainter companion to the west. Check out M73, cross it off your Messier list, and move on.

If you observe Aquarius' third Messier object next, you may think it's in a different class than M72 and M73. Globular cluster **M2** is a showpiece. To find it, scan roughly 4.5° due north of Beta Aquarii. If you have sharp eyes, you'll see this magnitude 6.6 cluster without optical aid from a dark site. A superb object in any telescope, M2 displays hundreds of stars through 10-inch and larger instruments.

Slightly more than 1° west of Nu Aquarii lies the **Saturn Nebula** (NGC 7009). Its name arises from the extensions, or ansae, at either end of the planetary nebula's disk that roughly resemble Saturn's rings. The extensions measure 15" past the ends of the 25"-long oval disk. At the end of the extensions are fainter bulbs you'll have trouble seeing through a 10-inch scope. Whether you see NGC 7009 as mainly blue or mainly green depends on your color perception.

While you're observing in Aquarius, don't miss the **Helix Nebula** (NGC 7293), which lies on Map 21. This magnitude 7.3 planetary nebula measures 13' across. Counteract its low surface brightness by using a nebula filter. With a filter in place, you'll see the ring structure through a telescope as small as 4 inches in aperture. Through a 12-inch scope, you may see slightly brighter concentrations on the north and south edges.

Designation	Right ascension	Declination	Magnitudes	Separation
Alpha² Capricorni	20h18m	−12°32'	3.8, 11.2	6.6"
Pi Capricorni	20h27m	−18°12'	5.2, 8.8	3.4"
Omicron Capricorni	20h30m	−18°35'	6.1, 6.6	18.9"
Gamma Delphini	20h47m	16°08'	4.5, 5.5	9.6"
4 Pegasi	21h39m	5°46'	5.8, 11.8	27.2"
30 Pegasi	22h20m	5°47'	5.4, 10.7	6.2"
34 Pegasi	22h27m	4°23'	5.8, 12.3	3.4"
Xi Pegasi	22h47m	12°10'	4.3, 12.3	11.5"
Tau¹ Aquarii	22h48m	−14°03'	5.8, 9.0	22.6"
94 Aquarii	23h19m	−13°27'	5.1, 7.5	12.6"
STF 3009	23h24m	3°42'	6.8, 8.8	7.0"
Omega² Aquarii	23h43m	−14°31'	4.6, 10.6	5.3"
107 Aquarii	23h46m	−18°40'	5.8, 6.8	6.6"

▲ **GLOBULAR CLUSTER** M15 in Pegasus is bright enough for you to glimpse with your naked eyes from a dark site. ADAM BLOCK/NOAO/AURA/NSF

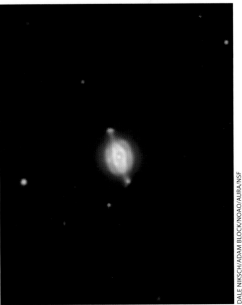

DALE NIKSCH/ADAM BLOCK/NOAO/AURA/NSF

▲ **THE SATURN NEBULA** (NGC 7009) was the first deep-sky object discovered by English astronomer William Herschel (1738–1822). He found it in 1782. NGC 7009 shines at 8th magnitude and measures 1.6' by 0.4'.

◄ **GLOBULAR CLUSTER** NGC 6934 in Delphinus spans 8.4'. While outlying stars are easy to resolve, you won't see many individual stars as you move toward the core. DANIEL VERSCHATSE

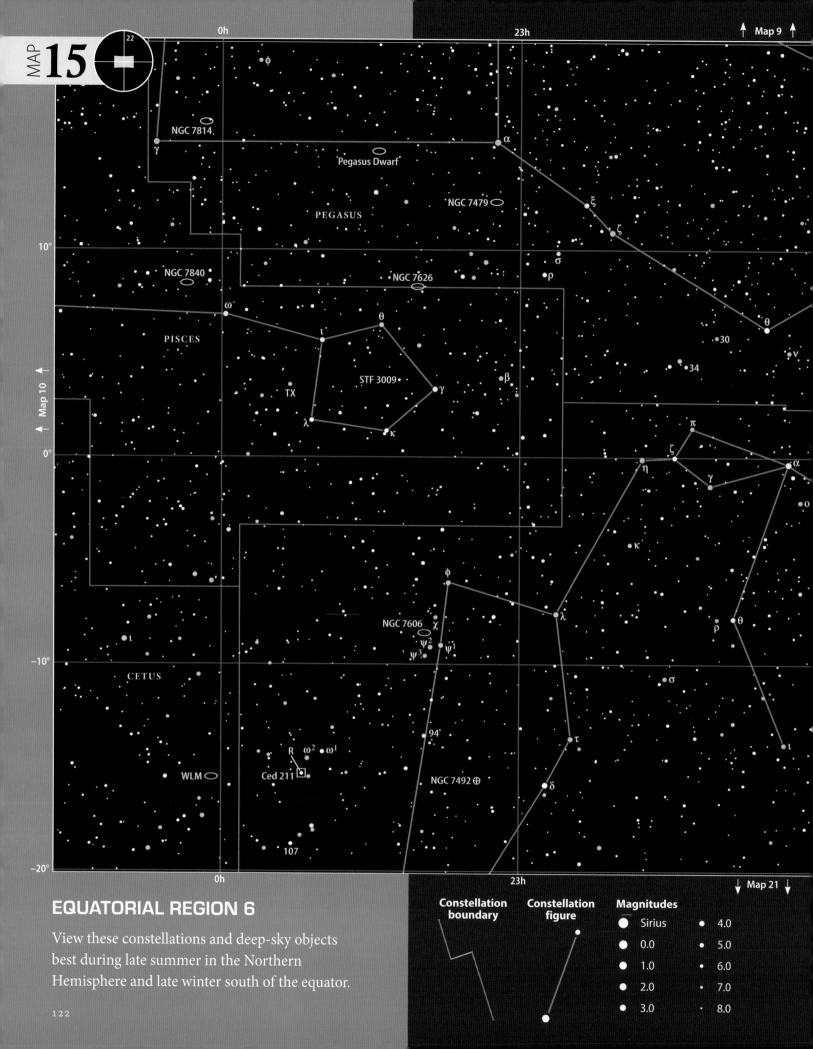

MAP

15

22

0h

23h

↑ Map 9 ↑

φ

NGC 7814

γ

Pegasus Dwarf

α

ξ

PEGASUS

NGC 7479

ζ

σ

NGC 7840

10°

ρ

NGC 7626

θ

ω

θ

30

PISCES

ι

34

θ

STF 3009

ν

TX

γ

β

λ

κ

π

ζ

0°

η

γ

α

ο

κ

φ

ρ

θ

NGC 7606

χ

λ

ψ²

ψ¹

ψ³

-10°

CETUS

σ

94

τ

R

ω²

ω¹

WLM

δ

Ced 211

NGC 7492 ⊕

ι

107

0h

23h

-20°

↓ Map 21 ↓

EQUATORIAL REGION 6

View these constellations and deep-sky objects
best during late summer in the Northern
Hemisphere and late winter south of the equator.

Constellation boundary	Constellation figure	Magnitudes	
		Sirius	4.0
		0.0	5.0
		1.0	6.0
		2.0	7.0
		3.0	8.0

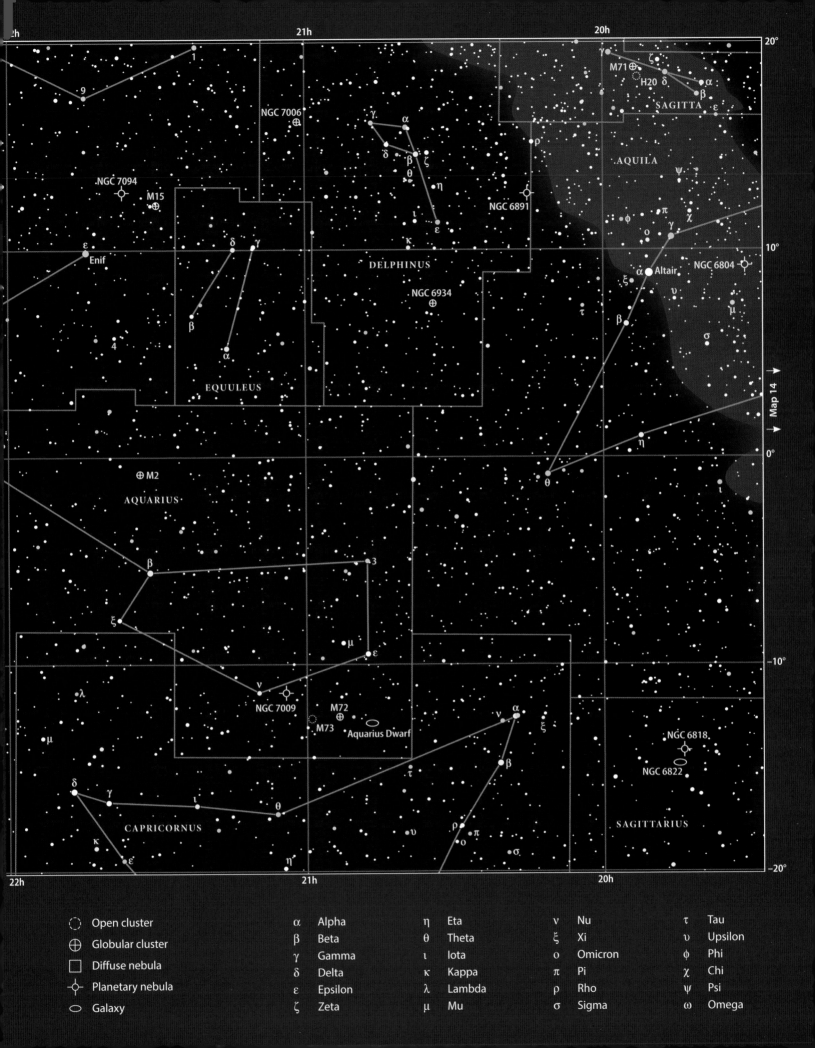

Open cluster
Globular cluster
Diffuse nebula
Planetary nebula
Galaxy

α	Alpha	η	Eta	ν	Nu	τ	Tau
β	Beta	θ	Theta	ξ	Xi	υ	Upsilon
γ	Gamma	ι	Iota	ο	Omicron	φ	Phi
δ	Delta	κ	Kappa	π	Pi	χ	Chi
ε	Epsilon	λ	Lambda	ρ	Rho	ψ	Psi
ζ	Zeta	μ	Mu	σ	Sigma	ω	Omega

MAP **16** SOUTH EQUATORIAL 1

◀ **THE SOUTHERN CIGAR GALAXY** (NGC 55) in Sculptor lies 5 million light-years away. Most of its stars lie to the west of its core. DSS/GIUSEPPE DONATIELLO

▼ **THE SOUTHERN PINWHEEL GALAXY** (NGC 300) in Sculptor resembles its northern namesake, M33 in Triangulum.

DSS/GIUSEPPE DONATIELLO

The sky's Furnace

As the star maps dip deeper into the southern sky, we start to see constellations unfamiliar to most Northern Hemisphere observers: Fornax the Furnace, Sculptor the Engraver's Tool, Phoenix the Phoenix, and Horologium the Clock. This region's stars are bright, but we're far from the Milky Way, which means galaxies abound.

For starters, insert an eyepiece that will give you at least a 2° field of view and point your telescope at the **Fornax Dwarf Galaxy**. Move your scope back and forth to bring out a haze just brighter than the background sky. This object doesn't have a corresponding NGC number because observers discovered it much later than 1888, when that catalog was published. This nearby galaxy — it's only 450,000 light-years away — is one of the Milky Way's closest dwarf companions.

If you're having trouble identifying the Fornax Dwarf, look for its brightest globular cluster, which is easier to see. NGC 1049 glows at magnitude 12.6 and is brighter toward its center.

Find the magnitude 4.5 star Beta Fornacis and move 2° north to the barred spiral galaxy **NGC 1097**. The core is this magnitude 9.5 galaxy's bright point, then the bar, which extends in a northwest-to-southeast direction, and, finally, the spiral arms. You'll need at least a 12-inch telescope and a dark sky to see this last feature.

Breaking from our galactic treasure hunt, we find planetary nebula **NGC 1360**. Its overall magnitude is a healthy 9.4, but this is spread out over a circle more than 6' across. As with all planetary nebulae, a nebula or OIII filter — which passes wavelengths planetaries emit — helps a lot.

The Virgo cluster of galaxies prominent in the spring sky is the sky's best-known galaxy cluster (see Map 13). Not far behind is the **Fornax galaxy cluster**. Under a dark sky, an 8-inch scope will let you see dozens of galaxies within a several-degree-wide swath of sky.

One of the brightest members of the Fornax cluster is magnitude 9.3 **NGC 1365**, the finest barred spiral galaxy in the sky. Its bar extends 4' in an east-west orientation; the central 2' is the nucleus. An 8-inch telescope easily resolves the spiral arms, of which the northern — extending from the west end of the bar — is the brightest.

The Fornax galaxy cluster's central region spans a scant 2°, but, through a 16-inch telescope, you'll see more than 100 galaxies. Even a 6-inch scope will show several dozen.

Start with NGC 1374. This magnitude 11 galaxy is 2.5' across and has a fainter companion (NGC 1375) 3' to the south. Pan ¼° southeast to reach NGC 1379, a circular elliptical galaxy that also glows at magnitude 11. Move north ½° to NGC 1380. This 10th-magnitude barred spiral shows a 4.5' by 2.5' oval with a 13th-magnitude star just to the west. Nearby, you'll notice magnitude 11.5 NGC 1381 and magnitude 10.8 NGC 1387.

Finally, in the same high-power field of view, you'll see the twin elliptical galaxies NGC 1399 and NGC 1404. Magnitude 8.8 NGC 1399 is the Fornax galaxy cluster's brightest member. It's not quite circular, measuring 6.9' by 6.5'. At magnitude 9.7, NGC 1404 is fainter and one-quarter the size of NGC 1399. NGC 1404 measures 3.3' by 3.0'.

Head south of Fornax into Eridanus and observe the barred spiral galaxy **NGC 1291**. You won't see the bar of this bright (magnitude 8.5) galaxy, but you will see its oval shape and intense core.

A bit more than 2° north of Tau⁴ Eridani lies **NGC 1300**, another classic barred spiral galaxy. The core of this magnitude 10.4 object appears oval. The spiral arms glow brightest near the ends of the nucleus' long axis, then wind tightly back around the bar.

Now, let's move into Sculptor. When you target objects in this constellation, you'll want to set aside a good portion of the night. You'll find no

▲ **THE SILVER COIN GALAXY** (NGC 253) in Sculptor shines at magnitude 7.6, bright enough to see with your unaided eyes under ideal conditions.

Designation	Right ascension	Declination	Magnitudes	Separation
HDO 183	0h46m	−47°33'	5.8, 13.5	14.3"
STN 60	1h05m	−33°31'	6.6, 10.6	8.6"
Epsilon Sculptoris	1h46m	−25°02'	5.5, 8.3	4.7"
Chi Eridani	1h56m	−51°36'	3.7, 10.7	4.8"
Omega Fornacis	2h34m	−28°13'	5.0, 7.7	10.9"
Eta² Fornacis	2h50m	−35°50'	5.8, 10.0	4.9"
Theta Eridani	2h58m	−40°18'	3.4, 4.5	8.3"
Tau⁴ Eridani	3h20m	−21°45'	4.0, 9.5	5.7"
B 1034	3h43m	−37°19'	4.6, 12.2	5.4"
Iota Phoenicis	23h35m	−42°36'	4.8, 12.8	6.7"
HWE 93	23h37m	−31°52'	6.5, 9.8	5.4"
Theta Phoenicis	23h40m	−46°38'	6.6, 7.2	3.9"
Delta Sculptoris	23h49m	−28°07'	4.6, 11.6	3.8"

▲ **NGC 1365** in Fornax rates as the sky's finest barred spiral galaxy. It shines at magnitude 9.3 and measures 8.9' by 6.5'.

less than four named galaxies in Sculptor, the highest number contained by any constellation. This quartet also is easy to see — each shines brighter than 9th magnitude.

You can find the **Southern Cigar Galaxy** (NGC 55) about 4° northwest of Alpha Phoenicis. This magnitude 8.1 galaxy lies 5 million light-years away. Nearly ½° long, NGC 55 is unusual because most of its stars are offset west of center, rather than concentrated in its core. This galaxy is one of the few that benefits from a nebula filter. Such a filter suppresses NGC 55's stars, and several large ionized hydrogen clouds pop into view.

Another magnitude 8.1 galaxy in Sculptor is the **Southern Pinwheel Galaxy** (NGC 300), which resembles its northern namesake, the Pinwheel Galaxy (M33) in Triangulum. Because it appears face-on, however, NGC 300 has a much lower surface brightness than NGC 55. To see the spiral arms well, you'll need a wide-field eyepiece on a 12-inch or larger scope.

The **Sculptor Dwarf Galaxy**, like the Fornax Dwarf, belongs to the Local Group of galaxies. Lying only 250,000 light-years away, the Sculptor Dwarf shines at magnitude 8.8. Don't let that brightness fool you, however. The Sculptor Dwarf stretches over 1.1° of sky, so its surface brightness is low. Use a wide-angle eyepiece in an 8-inch scope under a dark sky, and look for a diffuse haze just brighter than the background. It lies 2.3° south-southwest of magnitude 5.5 Sigma Sculptoris.

Sculptor's fourth named star system is the **Silver Coin Galaxy** (NGC 253), the showpiece deep-sky object in this region of sky. Observers with superb vision may see the magnitude 7.6 glow of this object with unaided eyes under perfect conditions and if NGC 253 is high enough in the sky. Prepare to spend some quality time at your telescope, and change eyepieces frequently to get the most out of observing this galaxy.

▲ **NGC 1398** in Fornax is a bright, large, tightly wound magnitude 9.7 spiral galaxy that measures 7.2' by 5.2'.

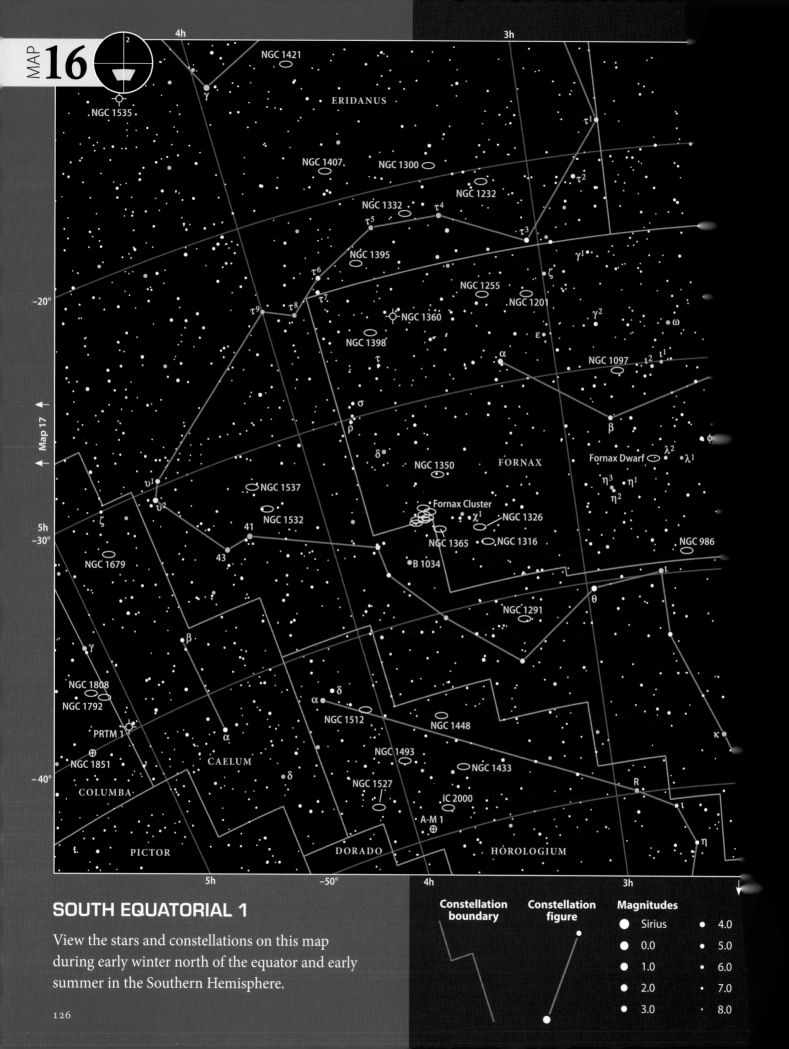

MAP
16

NGC 1421

NGC 1535

γ

ERIDANUS

τ¹

NGC 1407.   NGC 1300

τ²

NGC 1232

NGC 1332   τ⁴

τ⁵   τ³

γ¹

NGC 1395   ζ

NGC 1255

τ⁶   NGC 1201   γ²

ω

τ⁷   NGC 1360   ε

τ⁹   τ⁸   α   NGC 1097   ι² ι¹

NGC 1398

τ   β   φ

σ

ρ   Fornax Dwarf   λ² λ¹

δ   NGC 1350   FORNAX   η³ η¹

υ¹   NGC 1537   η²

υ²   Fornax Cluster

ζ   NGC 1532   χ¹   NGC 1326

41   NGC 1365   NGC 1316   NGC 986

43   B 1034   ι

NGC 1679   θ

NGC 1291

β   ι

γ

NGC 1808   δ

NGC 1792   α

PRTM 1   NGC 1512   NGC 1448   κ

NGC 1851   α   NGC 1493

CAELUM   NGC 1433   R

δ   NGC 1527   ι

COLUMBA   IC 2000   η

A-M 1

PICTOR   DORADO   HOROLOGIUM

SOUTH EQUATORIAL 1

View the stars and constellations on this map
during early winter north of the equator and early
summer in the Southern Hemisphere.

Constellation boundary	Constellation figure	Magnitudes	
		● Sirius	
		● 0.0	• 4.0
		● 1.0	• 5.0
		● 2.0	• 6.0
		● 3.0	• 7.0
			• 8.0

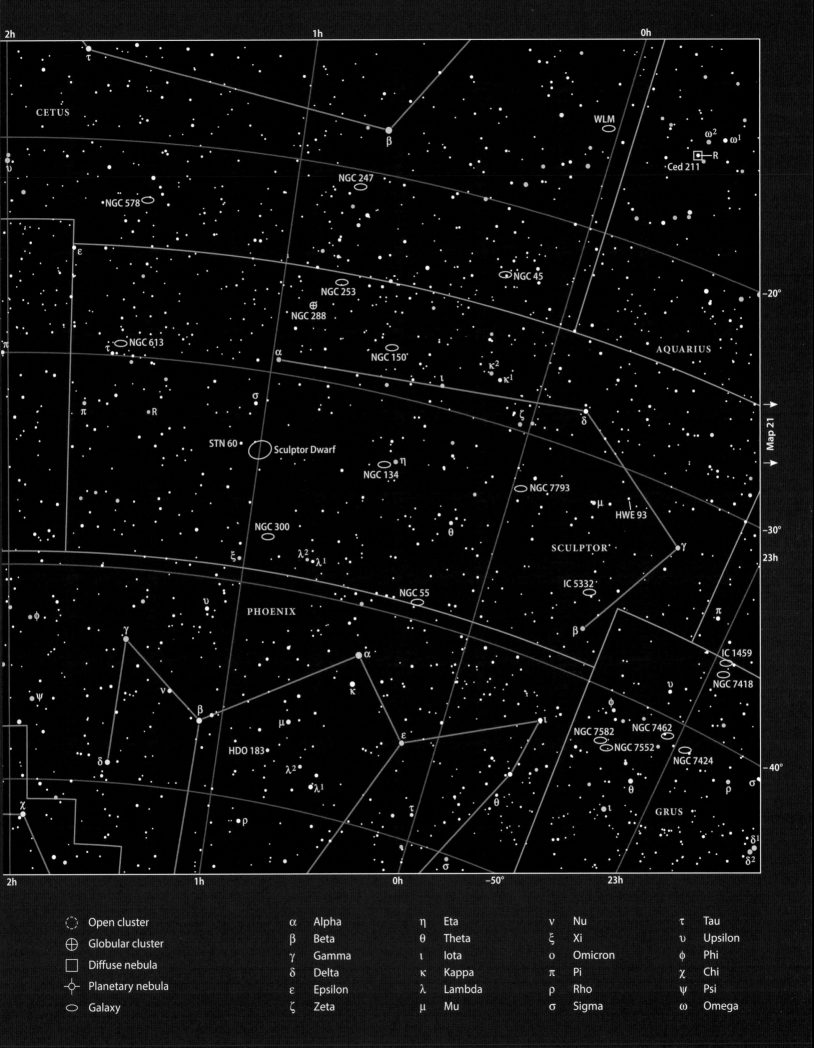

Open cluster ⊙ · ⊙
Globular cluster ⊕
Diffuse nebula ☐
Planetary nebula ⊹
Galaxy ⬭

α Alpha	η Eta	ν Nu	τ Tau
β Beta	θ Theta	ξ Xi	υ Upsilon
γ Gamma	ι Iota	ο Omicron	φ Phi
δ Delta	κ Kappa	π Pi	χ Chi
ε Epsilon	λ Lambda	ρ Rho	ψ Psi
ζ Zeta	μ Mu	σ Sigma	ω Omega

MAP **17** SOUTH EQUATORIAL 2

STEPHEN RAHN

▲ **OPEN CLUSTER** M41 in Canis Major is a naked-eye sight south of the night sky's brightest star, Sirius. The cluster measures 38' across.

South of the Dog Star

The night sky's two brightest stars — Sirius (Alpha Canis Majoris) and Canopus (Alpha Carinae) frame the northern and southern boundaries of the next star map. Between these two stars lie many superb deep-sky objects, especially galaxies and open clusters.

Most beginning observers know how to locate Sirius — draw a line down (toward the southeast) from Orion the Hunter's belt. If you continue that line for an equal distance past Sirius, you'll find yourself amid the stars of Puppis the Deck, most of which lies on Map 17.

Start 4° south of Sirius at **M41**, an open cluster you'll locate easily with unaided eyes. A 6-inch telescope reveals more than 50 stars across the ½° diameter of this magnitude 4.5 cluster.

An even brighter open cluster lies 8½° southeast of M41 — the **Tau Canis Majoris Cluster** (NGC 2362). Indeed, most of this cluster's magnitude 4.1 brightness comes from the star Tau Canis Majoris, which shines 3 magnitudes brighter than the group's next-brightest member. And when you look closely at Tau, you'll see a group of 10th-magnitude stars in a 6'-wide region surrounding it.

The brightest star cluster in Puppis isn't a star cluster at all. Through a telescope, **NGC 2451** appears as an impressive array of roughly a dozen bright stars around a reddish 4th-magnitude luminary. But because these stars are moving independently, and not as a group with a common center of gravity, NGC 2451 is just a chance alignment.

You'll find a true star cluster 1½° southeast of NGC 2451. At magnitude 5.8, open cluster **NGC 2477** is visible without optical aid to most observers from a dark site. So many stars of similar brightness crowd into an area ¼° across that NGC 2477 looks a bit like a loose globular cluster. Through a 6-inch scope, the central 5'-wide area resolves into 100 stars. Double your telescope's aperture to 12 inches, and you'll double the number of stars you can count. The 4th-magnitude star SAO 198545 just south of NGC 2477 lies in the foreground.

At the upper left of Map 17 is the bright planetary nebula **NGC 2440**. This magnitude 9.4 object looks slightly oval through a small telescope. A 12-inch scope shows the faint lobes at the northeast and southwest edges, as well as a faint haze surrounding the bright disk.

Lepus the Hare is a small constellation west of Canis Major that boasts a single Messier object — globular cluster **M79**. Although it glows at magnitude 7.7, this cluster is difficult to resolve in telescopes smaller than 8 inches. A larger scope will reveal many stars in M79's outer regions, as well as a large, densely packed core.

Look on Lepus' border with Canis Major to find **NGC 2196**. This galaxy is easy to find, even at magnitude 11.1, because it's small (2.8' by 2.2'), and therefore has a high surface brightness — well, at least the nucleus does. NGC 2196's spiral arms are faint and difficult to see because they wind closely around the nucleus.

When you're ready to observe in Columba the Dove, don't miss three standout deep-sky objects in its southwest corner: **NGC 1792**, **NGC 1808**,

▲ **THE TAU CANIS MAJORIS**
Cluster (NGC 2362) lies 5,000 light-years away and has a diameter of 6'. Most of the cluster's magnitude 4.1 light comes from its namesake star.

▲ **GLOBULAR CLUSTER** NGC 1851 in Columba has a core too dense to resolve visually. This cluster is an easy binocular object.

▲ **OPEN CLUSTER** NGC 2477 in Puppis shines at magnitude 5.8, bright enough for sharp-eyed observers to spot naked-eye. NGC 2477 measures 20' across.

◀ **GLOBULAR CLUSTER** M79 in Lepus the Hare shines at magnitude 7.7 from a distance of 42,000 light-years. TIM HUNTER

/// **DOUBLE-STAR DELIGHTS — MAP 17**

Designation	Right ascension	Declination	Magnitudes	Separation
DUN 16	3h49m	−37°37'	4.9, 5.4	8.0"
Alpha Caeli	4h41m	−41°51'	4.5, 12.5	6.6"
Gamma¹ Caeli	5h04m	−35°28'	4.7, 8.2	3.2"
DAW 117	5h21m	−34°21'	6.1, 10.9	2.2"
Alpha Columbae	5h40m	−34°04'	2.8, 12.5	13.5"
HJ 3869	6h33m	−32°01'	5.7, 7.7	24.9"
S 534	6h43m	−22°26'	6.3, 8.8	18.2"
Pi Canis Majoris	6h56m	−20°08'	4.6, 9.6	11.6"
Epsilon Canis Majoris	6h59m	−28°58'	1.6, 7.5	7.5"
HJ 3945	7h17m	−23°18'	4.8, 6.8	26.8"
Tau Canis Majoris	7h19m	−24°57'	4.4, 10.5	8.5"
Sigma Puppis	7h29m	−43°18'	3.3, 9.4	22.3"
HJ 4046	8h06m	−33°34'	6.0, 8.4	22.1"
HJ 4057	8h12m	−42°58'	4.9, 9.5	25.7"

and **NGC 1851**. The first two objects are galaxies that have interacted in the recent past. You'll need at least a 12-inch telescope just to see faint signs of the interaction. NGC 1808 also shows high star-forming activity.

NGC 1792's shape is not quite an oval. It's twice as long as wide with tightly wrapped spiral arms you'll need a big telescope to see. Notice how uniformly the magnitude 9.9 brightness spreads over NGC 1792's area.

Lying 40' to the northeast, NGC 1808 is a near twin of NGC 1792. NGC 1808 also shines at magnitude 9.9, and its dimensions are the same. Insert a low-power eyepiece, and you'll see both galaxies at once.

The third bright deep-sky object in this area is globular cluster NGC 1851. At magnitude 7.2, it's on the brink of naked-eye visibility. This cluster's outer stars resolve easily, but even at high magnification, the core remains too densely packed with stars to separate.

In the southwest (lower right) corner of Map 17, you can find two galaxies in the northern section of Horologium the Clock. **NGC 1512** makes an isosceles triangle with Alpha and Delta Horologii, lying roughly 2° from each. This magnitude 10.2 barred spiral galaxy has most of its brightness concentrated in its bar. A 16-inch scope reveals traces of the spiral arms, which extend from each end of the bar.

NGC 1527 lies 4½° south of NGC 1512. This magnitude 10.7 galaxy is twice as long (3') as wide. Look for two nearby stars: One shines at magnitude 12 and lies just north of the galaxy; the other glows at magnitude 13 and appears superimposed on NGC 1527's western region.

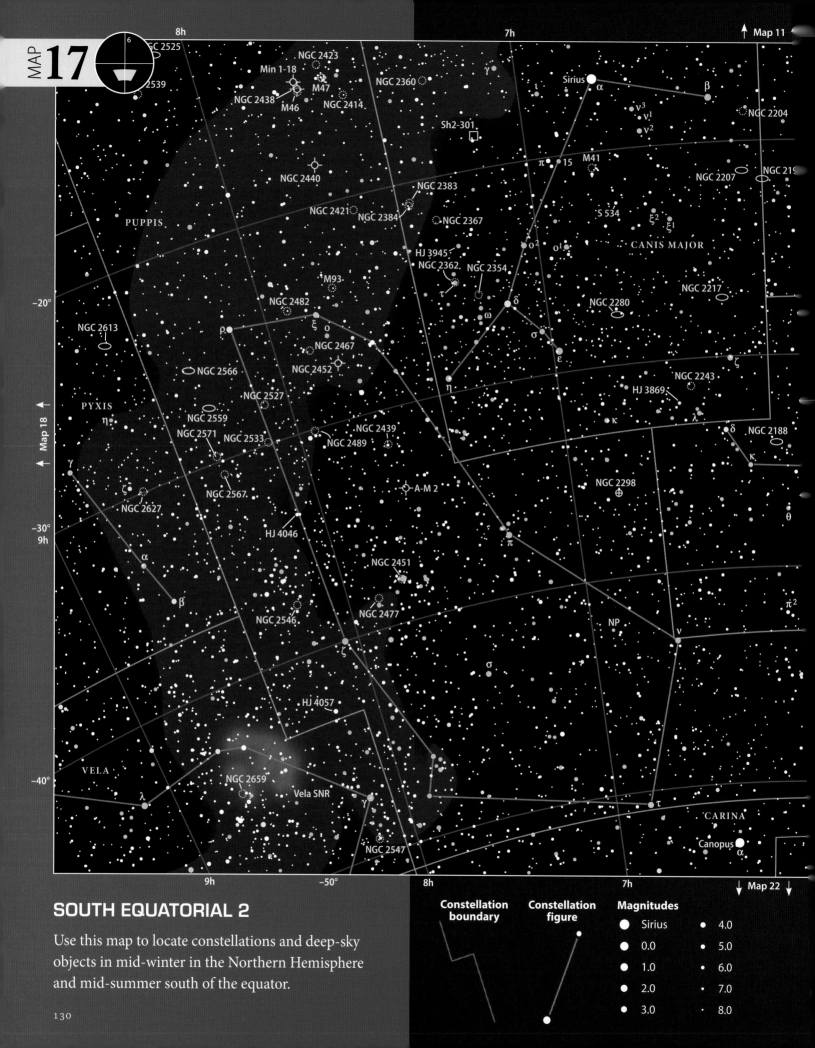

MAP 17

8h
7h
↑ Map 11

GC 2525
2539
NGC 2423
Min 1-18
M47
NGC 2438
M46
NGC 2414
NGC 2360
NGC 2204
γ
Sirius
α
β

NGC 2440
Sh2-301
π
15
M41
ν³
ν¹
ν²
NGC 2207
NGC 219

PUPPIS
NGC 2421
NGC 2383
NGC 2384
NGC 2367
S 534
ξ²
ξ¹
CANIS MAJOR

−20°
HJ 3945
NGC 2362
NGC 2354
o²
o¹
NGC 2280
NGC 2217

NGC 2613
M93
NGC 2482
τ
ω
δ

ρ
ξ
o
NGC 2467
σ
ε
ζ

NGC 2566
NGC 2452
NGC 2527
η
NGC 2243
HJ 3869

PYXIS
η
NGC 2559
NGC 2571
NGC 2533
NGC 2439
NGC 2489
κ
λ

Map 18
γ
NGC 2627
ζ
NGC 2567
A-M 2
NGC 2298
δ
NGC 2188
κ

−30°
9h
HJ 4046
π
θ

α
NGC 2451

β
NGC 2477
NGC 2546
NP
ν
π²

−40°
HJ 4057
σ

VELA
NGC 2659
Vela SNR
γ
τ
CARINA

λ
NGC 2547
Canopus
α

9h
−50°
8h
7h
↓ Map 22 ↓

SOUTH EQUATORIAL 2

Use this map to locate constellations and deep-sky
objects in mid-winter in the Northern Hemisphere
and mid-summer south of the equator.

Constellation boundary	Constellation figure	Magnitudes	
		Sirius	4.0
		0.0	5.0
		1.0	6.0
		2.0	7.0
		3.0	8.0

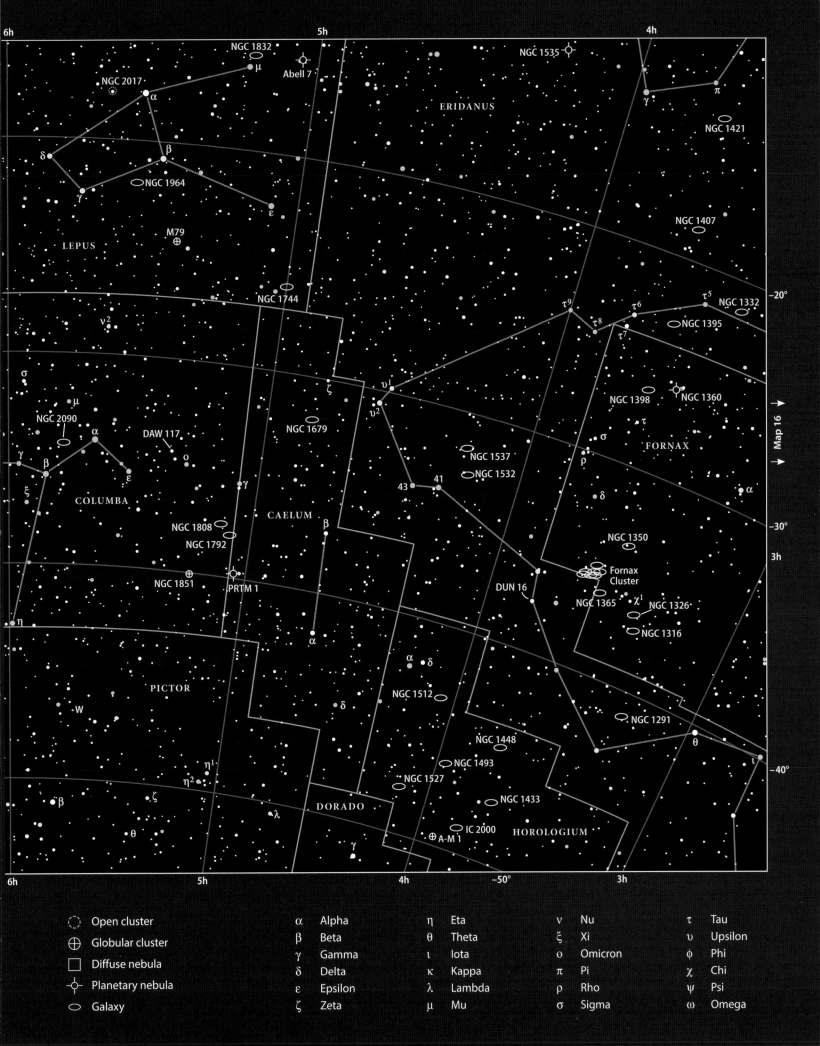

NGC 1832
μ
NGC 2017
α
Abell 7
NGC 1535
ERIDANUS
γ
π
NGC 1421
δ
β
NGC 1964
γ
ε
NGC 1407
M79
LEPUS
NGC 1744
-20°
ν²
τ⁹
τ⁶
τ⁵
NGC 1332
τ⁸
NGC 1395
σ
τ⁷
υ¹
NGC 1398
NGC 1360
μ
NGC 2090
υ²
τ
NGC 1679
FORNAX
α
DAW 117
σ
NGC 1537
ρ
γ
β
ο
NGC 1532
δ
ε
43
41
α
ξ
COLUMBA
CAELUM
NGC 1808
β
-30°
NGC 1792
γ
NGC 1350
3h
NGC 1851
Fornax
Cluster
PRTM 1
DUN 16
NGC 1365
χ¹
NGC 1326
η
NGC 1316
PICTOR
NGC 1291
W
θ
α
δ
ι
NGC 1512
-40°
δ
NGC 1448
η¹
NGC 1493
η²
NGC 1527
NGC 1433
β
ζ
θ
λ
DORADO
IC 2000
HOROLOGIUM
γ
A-M 1

Map 16

⟡	Open cluster		α	Alpha		η	Eta		ν	Nu		τ	Tau
⊕	Globular cluster		β	Beta		θ	Theta		ξ	Xi		υ	Upsilon
□	Diffuse nebula		γ	Gamma		ι	Iota		ο	Omicron		φ	Phi
✧	Planetary nebula		δ	Delta		κ	Kappa		π	Pi		χ	Chi
⬭	Galaxy		ε	Epsilon		λ	Lambda		ρ	Rho		ψ	Psi
			ζ	Zeta		μ	Mu		σ	Sigma		ω	Omega

MAP 18 · SOUTH EQUATORIAL 3

▲ **THE PENCIL NEBULA** (NGC 2736) makes up the east-southeast part of the Vela Supernova Remnant, which originated less than 10,000 years ago.

Spring's Water Snake

The southern portion of Hydra — the sky's largest constellation — dominates the top part of the next star map. Because it is so large, Hydra boasts a variety of celestial gems.

Take a break from observing traditional deep-sky objects to spot **V Hydrae**, a single star many observers rate as the reddest in the sky. V Hydrae is a variable star with a period (from one peak brightness to the next) of 531 days. If it's at its maximum magnitude of 6.6, you may be able to spot it with unaided eyes from a dark site. At minimum, the star's brightness drops to magnitude 9.0. Its color, however, gives it away. Use a telescope, and slightly defocus the image to see V Hydrae at its reddest.

A bright galaxy in Hydra, **NGC 3621**, shines at magnitude 8.9. Twice as long as it is wide (9.8' by 4.6'), NGC 3621 has a bright nucleus that covers only 15". NGC 3621 is a spiral galaxy, but you'll need a 14-inch or larger scope to see any of its structure.

Not many planetary nebulae shine as brightly as magnitude 7.8, so when you have a chance to observe one, don't miss it. Such an object, the **Ghost of Jupiter** (NGC 3242), lies 2° south of Mu Hydrae. Objects like

the Ghost of Jupiter inspired English astronomer John Herschel to coin the term "planetary nebulae." NGC 3242 measures 13" across, which is about the same as Mars at a distant opposition (the point in its orbit when it lies opposite the Sun in our sky). At low magnification, the Ghost's color — blue-green — resembles Uranus'. Once you've verified this, crank up the power in steps. You'll see a faint shell 40" across surrounding a brighter, football-shaped interior. An even closer look at the object reveals an empty space 10" across that contains the planetary's central star.

The constellation due south of the Ghost of Jupiter is Antlia the Air Pump. Small and sparse, Antlia contains only four stars brighter than 5th magnitude. Another slightly fainter star you'll spot easily from a dark site is **U Antliae**. This star is similar in color to V Hydrae, described above, only brighter. Such objects are known as carbon stars, and they're all intensely red. For a list of the sky's reddest stars, see page 62.

Magnitude 9.3 **NGC 2997** is one of only two galaxies in Antlia that shows any detail through medium-sized telescopes. Through a 10-inch scope, you'll see an indistinct glow 5' by 3' aligned east-west. The other galaxy is **NGC 3175**, a nearly edge-on spiral that is three times as long as

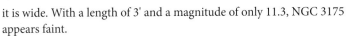
NICHOLAS JONES

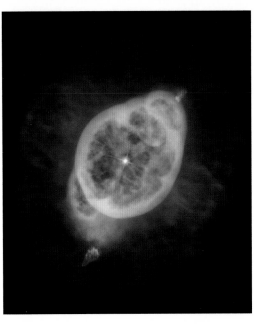

◀ **THE GHOST OF JUPITER** (NGC 3242) is a spherical planetary nebula whose color more closely resembles the blue-green of Uranus. JUDY SCHMIDT

Designation	Right ascension	Declination	Magnitudes	Separation
DUN 70	8h30m	−44°43'	5.2, 6.8	4.5"
Eta Pyxidis	8h38m	−26°15'	5.2, 13.0	16.0"
Beta Pyxidis	8h40m	−35°18'	4.0, 12.5	12.7"
COO 74	8h40m	−40°15'	5.2, 8.5	4.0"
Epsilon Pyxidis	9h10m	−30°21'	5.6, 9.9	17.7"
Zeta¹ Antliae	9h31m	−31°53'	6.3, 7.2	8.0"
HJ 4218	9h33m	−36°24'	7.6, 10.5	5.7"
SHJ 110	10h04m	−18°06'	5.8, 8.0	21.2"
Delta Antliae	10h30m	−30°36'	5.7, 9.7	11.0"
Struve 1474	10h48m	−15°14'	7.5, 7.8	6.8"
BSO 6	11h29m	−42°40'	5.4, 8.1	13.0"
HJ 4455	11h37m	−33°34'	6.0, 8.1	3.3"
HJ 4518	12h25m	−41°23'	6.2, 9.5	10.0"

it is wide. With a length of 3' and a magnitude of only 11.3, NGC 3175 appears faint.

It's worth stopping in Pyxis to view two open star clusters. The first is magnitude 8.4 **NGC 2627**. If you use an 8-inch scope at 50x, you'll see 40 stars in a region 6' across and many more hinted at in the background.

Now move to **NGC 2818**. At magnitude 8.2, it's a bit brighter than NGC 2627, but it's special for a different reason: NGC 2818 contains a planetary nebula. The nebula looks like a small version of the Dumbbell Nebula (M27) in Vulpecula. You'll see 30 stars in an area about 9' across.

Vela is a large constellation with few bright stars. In the 18th century, the constellation Vela did not exist. Its stars were part of a larger constellation named Argo Navis, which, according to Greek mythology, was the ship Jason and the Argonauts used to find the Golden Fleece. The constellation Argo Navis was so unwieldy that later astronomers subdivided it into Carina the Keel, Puppis the Deck, and Vela the Sails.

The **Vela Supernova Remnant** is the sky's largest supernova remnant, covering 5°. The best way to observe this object is to use a 12-inch or larger telescope with a low-power eyepiece and a nebula filter, such as an OIII. Either disengage the drive motor, or set your slewing speed at "medium," and scan the area. Only one portion of this region has an NGC number assigned to it. The east-southeast region carries the designation **NGC 2736**, and observers most often call it the Pencil Nebula.

Look 2° south of Gamma Velorum for **NGC 2547**. At magnitude 4.7, it's an easy naked-eye object. NGC 2547 measures 1° across, but through a 6-inch or larger scope, you'll see nearly all of its stars in a ½° area. And, yes, the magnitude 6.5 star is part of the cluster, not a foreground star.

Get out your binoculars (and Map 23) and look 2° north-northwest of Delta Velorum for **IC 2391**, a magnitude 2.5 open cluster almost 1° across. Binoculars highlight this object's brightest stars, but a telescope with a wide-angle eyepiece will let you go a bit deeper. In all, expect to see about 30 stars brighter than 12th magnitude.

The **Eight-Burst Nebula** (NGC 3132) lies in Vela's northeast corner. This is a magnitude 9.7 planetary nebula comprising a 10th-magnitude central star surrounded by a shell measuring 60" by 45". The shell's outer regions are brightest, but a 10-inch or larger telescope will allow you to see this object's splotchy interior.

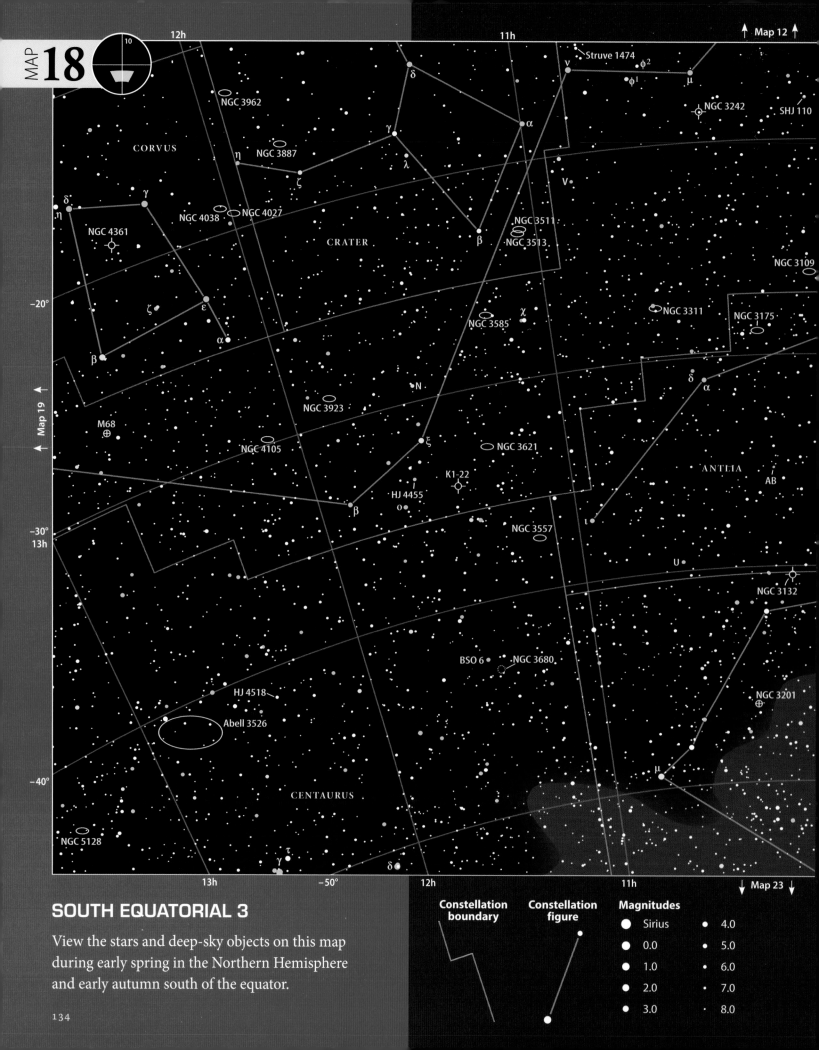

MAP 18

↑ Map 12 ↑

12h

11h

Struve 1474

ν

φ²

φ¹

μ

SHJ 110

NGC 3962

δ

NGC 3242

CORVUS

γ

NGC 3887

η

λ

V

ζ

α

NGC 3511

NGC 3109

δ

γ

NGC 4038

NGC 4027

β

NGC 3513

η

CRATER

NGC 4361

χ

NGC 3311

NGC 3175

−20°

ζ

ε

NGC 3585

β

α

N

δ

α

Map 19

M68

NGC 3923

ANTLIA

NGC 4105

ξ

NGC 3621

AB

K1-22

HJ 4455

β

o

NGC 3557

ι

−30°

13h

U

NGC 3132

HJ 4518

BSO 6

NGC 3680

NGC 3201

Abell 3526

μ

−40°

CENTAURUS

NGC 5128

τ

γ

δ

13h

−50°

12h

11h

↓ Map 23 ↓

SOUTH EQUATORIAL 3

View the stars and deep-sky objects on this map
during early spring in the Northern Hemisphere
and early autumn south of the equator.

Constellation boundary	Constellation figure	Magnitudes	
		⬤ Sirius	• 4.0
		⬤ 0.0	• 5.0
		⬤ 1.0	• 6.0
		⬤ 2.0	· 7.0
		• 3.0	· 8.0

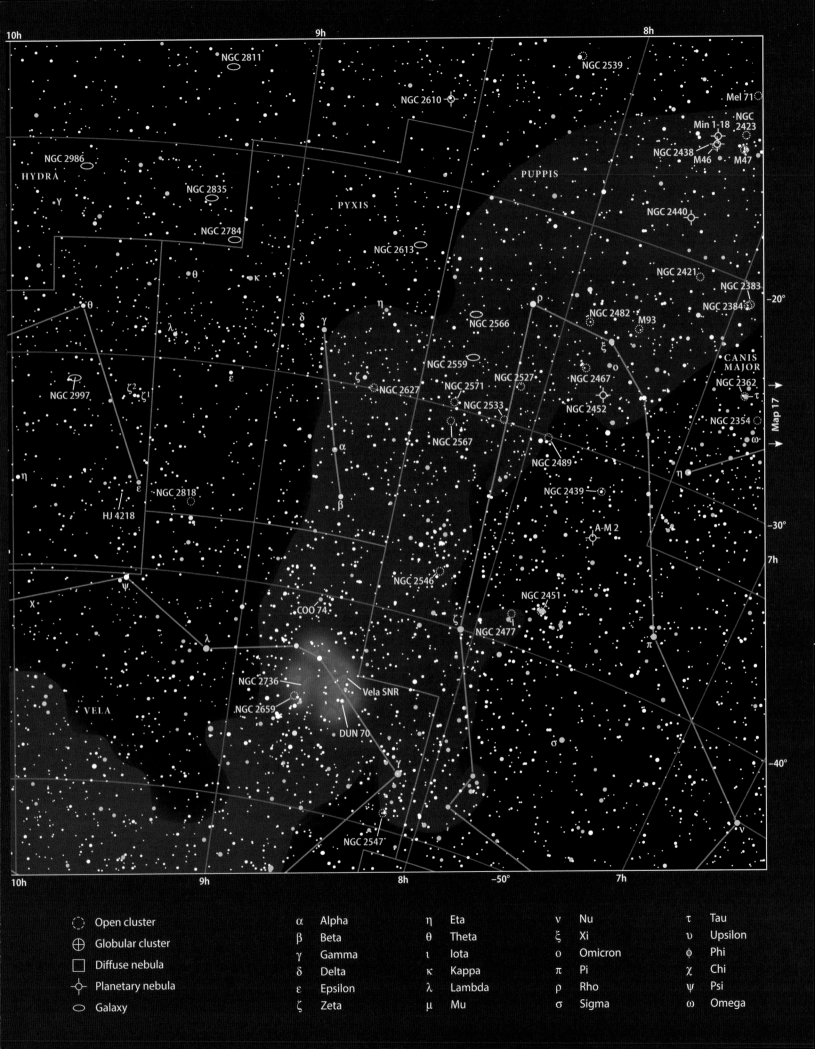

NGC 2811

NGC 2539

Mel 71

NGC 2610

NGC 2423
Min 1-18
NGC 2438
M46   M47

NGC 2986

HYDRA

PUPPIS

γ

NGC 2835

NGC 2440

PYXIS

NGC 2784

NGC 2421

NGC 2383

NGC 2613

NGC 2384

−20°

θ

κ

η

NGC 2566

ρ

CANIS
MAJOR

δ   γ

NGC 2559

NGC 2482   M93

NGC 2362   τ

θ

ξ   o

λ

ζ

NGC 2527

NGC 2467

NGC 2354

ε

NGC 2627

NGC 2571

NGC 2452

ω

NGC 2997

ζ² ζ¹

NGC 2533

α

NGC 2567

NGC 2489

η   Map 17

η

NGC 2439

−30°

ε

NGC 2818

A-M 2

HJ 4218

β

7h

ψ

NGC 2546

NGC 2451

χ

NGC 2477

λ

ζ

NGC 2477

π

COO 74

σ

NGC 2736

Vela SNR

VELA

NGC 2659

DUN 70

−40°

γ

ν

NGC 2547

10h   9h   8h   −50°   7h

MAP **19** SOUTH EQUATORIAL 4

▲ **GLOBULAR CLUSTER** M68 in Hydra lies 33,000 light-years from Earth. This object measures 12' across and shines at magnitude 7.6.

Wolf and Centaur

Two large, bright constellations — Lupus the Wolf and Centaurus the Centaur — dominate the next star map. The majority of each constella-tion lies along the Milky Way's border, so a variety of deep-sky objects awaits your inspection. Hydra's faint eastern half also appears.

Start 2.9° east-northeast of magnitude 3.4 Zeta Lupi with **NGC 5927**, a magnitude 8.0 globular cluster that's 12' across. You may need a telescope as large as 16 inches to resolve this cluster's myriad faint stars.

Globular cluster **NGC 5986** — at magnitude 7.5 — is a bit brighter than NGC 5927, and its stars resolve more easily. A 6-inch scope shows a few stars. Move up to a 12-inch under a dark sky, crank the magnification up to 200x, and dozens more stars will pop into view. NGC 5986 makes a small triangle with a 6th- and 7th-magnitude star ¼° to the east.

Two planetary nebulae — **IC 4406** and **NGC 5882** — deserve some of your observing time. IC 4406 glows at magnitude 10.2 and is roughly 1.5' across. An 8-inch or larger telescope reveals a rectangular object much brighter at its center with fainter extensions stretching east and west.

NGC 5882 shines at magnitude 9.4 and stretches 18" in diameter. It's circular and glows a vivid blue-green through an 8-inch or larger scope.

A visual survey of Centaurus reveals a fuzzy star — Omega — that happens to be the sky's brightest globular cluster, **NGC 5139**. Gleaming at magnitude 3.5, this object was misidentified by German stellar mapmaker

Johannes Bayer. In his 1603 work, *Uranometria* — the first star maps to use Greek letters to identify the stars — Bayer assigned this globular the Greek letter Omega, a designation it has retained ever since.

Omega Centauri is a wonder to behold through binoculars or telescopes of any size. Omega appears slightly larger than the Full Moon, and, because it's rotating relatively quickly, its shape is slightly out-of-round. Through an 8-inch telescope, you'll see 1,000 stars, each a faint pinprick of light. At high power, the stars appear nearly uniformly distributed across the field of view.

Move 4.5° north of NGC 5139 to find irregular galaxy **Centaurus A** (NGC 5128). Although this object shines at magnitude 6.7, you won't see it with unaided eyes even from the darkest site because its light spreads over an area that measures 31' by 23'. Through a 12-inch telescope, you'll see the central circular haze 8' across divided by a dark lane 1' wide. A 12th-magnitude star lies in front of the dust lane at the northwest end.

Only 17' from Iota Centauri sits spiral galaxy **NGC 5102**, which measures 10' by 4' and shines at magnitude 8.8. Most of NGC 5102's brightness comes from its core. Be sure Iota Cen lies outside the field of view when you observe this galaxy, or the glare will overwhelm the view.

For something a bit different, look 2° north of Phi Centauri. This object is the bright reflection nebula **NGC 5367**. Through a 12-inch telescope, you'll see an evenly illuminated haze roughly 2' across. To the northeast is a detached region that measures 2' by 1'.

◀ NGC 5078 in Hydra looks like a sandwich because of its dark lane. Below is its companion galaxy, IC 879. DANIEL VERSCHATSE

CHRISTIAN REUSCH

▲ BARRED SPIRAL GALAXY M83 appears face-on, an alignment that makes this magnitude 7.5 galaxy a showpiece.

Designation	Right ascension	Declination	Magnitudes	Separation
JC 17	12h10m	−34°42'	6.3, 8.1	3.4"
Xi² Centauri	13h07m	−49°54'	4.4, 9.5	25.1"
HWE 94	13h49m	−35°41'	6.6, 9.6	11.6"
RMK 18	13h52m	−52°48'	5.7, 7.9	18.1"
4 Centauri	13h53m	−31°55'	4.8, 8.5	14.9"
54 Hydrae	14h46m	−25°26'	5.2, 7.2	8.4"
Kappa Centauri	14h59m	−42°06'	3.4, 11.5	3.9"
HJ 4727	15h04m	−27°50'	8.6, 8.7	7.6"
Kappa¹ Lupi	15h12m	−48°44'	4.1, 6.0	26.6"
Upsilon Librae	15h37m	−28°07'	3.8, 10.8	3.3"
Omega Lupi	15h38m	−42°33'	4.3, 11.0	11.8"
Xi¹ Lupi	15h57m	−33°57'	5.3, 5.8	10.4"
Eta Lupi	16h00m	−38°24'	3.6, 7.8	15.0"

Now, move to the eastern part of Hydra, which also lies on this star map. Hydra's brightest galaxy, and a showpiece of the "near-southern" sky, is **M83** (NGC 5236). M83 is a barred spiral galaxy that appears nearly face-on. This fortuitous alignment lets us observe its spiral arms through telescopes as small as 6 inches in diameter. M83's bar is aligned northeast-to-southwest. The galaxy's core is brilliant and unresolvable. Through a 12-inch or larger scope, look for dark lanes of dust and cold gas within the spiral arms. M83 shines at magnitude 7.5 and measures 15' long.

Another Messier object in southern Hydra is globular cluster **M68** (NGC 4590). At magnitude 7.6, this object glows slightly fainter than M83. The star cluster looks brighter, however, because its light spreads over an area only 12' across. Although it's bright, M68 doesn't resolve well. Even through a 12-inch scope, you'll see fewer than 30 individual stars in front of the cluster's nucleus.

◀ OMEGA CENTAURI (NGC 5139) outshines every other globular cluster in the sky. Through any size telescope, this object is a wonder to behold.

STEPHEN RAHN

MAP
19
14
16h
15h
↑ Map 13 ↑

SOUTH EQUATORIAL 4

View the stars and constellations on this map
during mid-spring north of the equator and
mid-autumn in the Southern Hemisphere.

138

Constellation boundary

Constellation figure

Magnitudes

● Sirius	● 4.0	
● 0.0	● 5.0	
● 1.0	● 6.0	
● 2.0	· 7.0	
● 3.0	· 8.0	

MAP 20 · SOUTH EQUATORIAL 5

MARC VAN NORDEN

▲ **GLOBULAR CLUSTER** M4 in Scorpius lies near the bright reddish star Antares (not shown). M4 is visible easily with the naked eye.

Designation	Right ascension	Declination	Magnitudes	Separation
HJ 4788	15h36m	−44°57'	5.0, 7.0	2.1"
2 Scorpii	15h54m	−25°19'	4.7, 7.4	2.3"
Beta Scorpii	16h05m	−19°48'	2.9, 6.9	13.6"
Sigma Scorpii	16h21m	−25°35'	2.9, 8.7	20.0"
Rho Ophiuchi	16h26m	−23°26'	5.2, 5.9	3.1"
Alpha Scorpii	16h29m	−26°26'	1.2, 5.4	2.9"
Omicron Ophiuchi	17h18m	−24°17'	5.4, 6.9	10.2"
Xi Ophiuchi	17h21m	−21°06'	4.5, 9.0	3.9"
Eta Sagittarii	18h18m	−36°44'	3.2, 7.8	3.6"
Beta¹ Sagittarii	19h23m	−44°27'	4.3, 7.4	28.3"
52 Sagittarii	19h37m	−24°52'	4.7, 9.2	2.6"

Our galaxy's heart

The following star map will keep your telescope's drive humming through many observing sessions. When we look toward this sky region, we face our Milky Way's center, where star clusters and nebulae abound. In Sagittarius alone, we find 15 Messier objects — the largest number in any of the 88 constellations. Scorpius adds four more. We'll start our tour, however, in the small constellation of Norma the Carpenter's Square.

Several nice open star clusters populate the southeastern part of Norma, which lies at the bottom-right of Map 20. Norma continues to Map 24, where we find **NGC 6067**, a magnitude 5.6 naked-eye gem that displays about 100 stars when viewed through a 6-inch telescope. Adding to the appeal of this object is its placement within the Norma Star Cloud, a rich region of the Milky Way. Scopes above 14 inches in aperture will reveal several hundred additional stars.

Norma's other naked-eye open cluster, **NGC 6087** (Map 24), shines at magnitude 5.4 and measures 15' across. Most of its light comes from the variable star S Normae. Every 9.75 days, this star's magnitude varies between 6.1 and 6.8. NGC 6087 contains more than 50 stars visible through 8-inch telescopes, but the field is so crowded with faint background stars that — except for a dozen brighter members — you'll have trouble determining which stars belong to the cluster.

Corona Australis the Southern Crown contains a bright globular cluster, **NGC 6541**. You can see this cluster, which glows at magnitude 6.3, without optical aid if it's high enough in the sky. A 4-inch telescope gives a great view, but, through a 12-inch, you'll see more than 100 outlying stars around a concentrated core.

IC 1297, a magnitude 10.7 planetary nebula, lies 1.5° east of Beta Coronae Australis. You'll need to crank up the magnification on this

7"-diameter object because it appears stellar at low powers. A nebula filter helps because it allows light from the planetary through, but it dims the light from surrounding stars. IC 1297's faint bluish hue is difficult to see in telescopes with apertures less than 16 inches.

Telescopium contains a small cluster of galaxies, and the group makes a good target for a large telescope. The brightest member, **NGC 6868**, glows at magnitude 10.6. Other cluster galaxies include NGC 6861 (26' to the west) and NGC 6870 (7' to the north), but both are several magnitudes fainter than NGC 6868.

In Scorpius, you'll encounter a wealth of deep-sky objects, including four Messier objects: **M4**, **M6**, **M7**, and **M80**. This constellation offers so much more, however. Look 1° northeast of M4, and you'll find another nice globular cluster, although not nearly as bright. **NGC 6144** shines at magnitude 9.0 and measures 9.3' across, about one-third M4's diameter. You'll only see it to about half this diameter, however, unless you use a 12-inch or larger telescope. Be sure to move Antares out of the field of view.

Scorpius is awash in planetary nebulae. Among the best are **NGC 6072** (magnitude 11.7, 40" across), **NGC 6153** (magnitude 10.9, 25" across), and **NGC 6337** (magnitude 12.3, 48" across). By far the best planetary in Scorpius, however, is the **Bug Nebula** (NGC 6302).

Find the Bug surrounded by the Scorpion's tail midway between Lambda and Mu Scorpii. NGC 6302 measures 2' by 1'. A prominent lobe with a tapered end makes up the nebula's western side. A faint extension protrudes from its eastern edge. At magnitude 9.6, NGC 6302 is a bright planetary; use an OIII filter to bring out its subtle details.

Under a dark sky, it would be hard to miss the Scorpius OB1 association, a huge group of hot stars. It begins at Zeta Scorpii and extends northward nearly halfway to Mu Scorpii. The brightest contributor — open cluster **NGC 6231** — lies at the southern end only 32' from Zeta. This magnitude 2.6 gem spans 14' and displays more than 100 stars through 6-inch and larger telescopes. The tight grouping of stars at the cluster's center is particularly striking.

Two beautiful emission nebulae, each measuring roughly ½° across, are must-see objects. First, find the star-forming region known as the **Cat's Paw Nebula** (NGC 6334). The Cat's Paw actually comprises five individual nebulous patches. The one at the southeastern end is brightest.

After you've marveled at the Cat's Paw Nebula, move about 1.8°

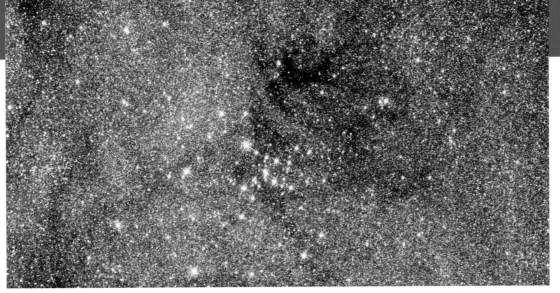

▲ **OPEN CLUSTER** M7 in Scorpius lies near the Scorpion's stinger. ALLAN COOK/ADAM BLOCK/NOAO/AURA/NSF

▲ **THE BUG NEBULA (NGC 6302)** in Scorpius is a strange-looking planetary nebula in which the old star's gas expanded unevenly.

ADAM BLOCK/NOAO/AURA/NSF

/// MESSIER OBJECTS IN SAGITTARIUS

Object	Type	Magnitude	Size
M8	N	6.0	45' by 30'
M17	N	7.0	20' by 15'
M18	OC	6.9	10'
M20	N	9.0	20'
M21	OC	5.9	13'
M22	GC	5.2	24'
M23	OC	5.5	27'
M24	SC	2.5	95' by 35'
M25	OC	4.6	32'
M28	GC	6.9	11.2'
M54	GC	7.2	9.1'
M55	GC	6.3	19'
M69	GC	7.4	7.1'
M70	GC	7.8	7.8'
M75	GC	8.6	6'

GC = globular cluster; N = bright nebula; OC = open cluster; SC = star cloud

▲ **THE CAT'S PAW NEBULA (NGC 6334,** lower right) complements NGC 6357, which lies less than 2° to its north-northeast.

▶ **M24** (top) is a bright 1.6°-long section of the Sagittarius Star Cloud. This Messier object is an asterism, not a star cluster. M24 lies 10,000 light-years away.

DANIEL VERSCHATSE

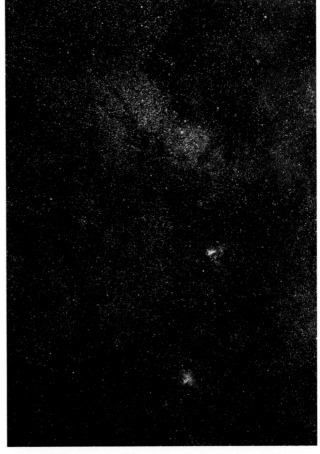

STEPHEN RAHN

north-northeast to **NGC 6357**. While you'll see the whole extent of the Cat's Paw, only the west-central region of NGC 6357 — involved with a small star cluster — shines brightly enough for medium-sized scopes.

Finally, we arrive at Sagittarius the Archer. For many Northern Hemisphere observers, Sagittarius lies too near the horizon to provide a quality observing experience. For those who live (or can travel to) where the constellation is high in the sky, even a whole season of observing won't be nearly enough to cover all of its deep-sky treats.

In addition to the Messier objects, Sagittarius contains a score of bright deep-sky gems, mainly nebulae and star clusters. One extra galactic target you can find is **Barnard's Galaxy** (NGC 6822). This object is somewhat of an observing challenge. It glows at a respectable magnitude of 8.8, but its light spreads over an area 19' by 15'. This combination means it's a low-surface-brightness object, so you'll need a dark sky to see it. Because NGC 6822 lies only 1.6 million light-years away, you can use a nebula filter with an 8-inch or larger telescope to see HII regions — vast clouds of glowing hydrogen that eventually will form stars.

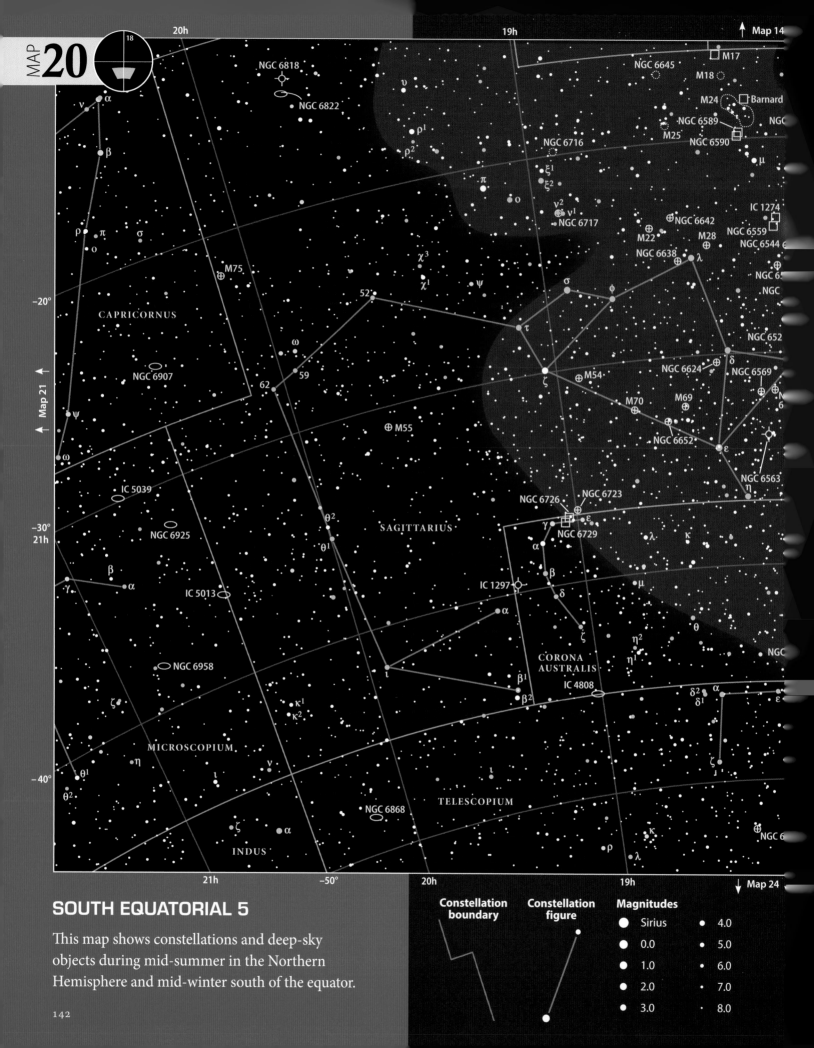

MAP 20
18
20h
19h
↑ Map 14

NGC 6818

M17

NGC 6645

M18

NGC 6822

M24   Barnard

NGC

NGC 6589

M25

NGC 6590

υ

ρ¹

ρ²

NGC 6716

μ

ν   α

β

ξ¹

ξ²

π

o

ν²   ν¹

NGC 6717

ρ   π

σ

χ³

χ¹

ψ

IC 1274

NGC 6642

NGC 6559

o

M22

M28

NGC 6544

M75

σ

φ

λ

NGC 6638

NGC 6

−20°

CAPRICORNUS

52

τ

NGC 652

ω

δ

59

ζ

M54

NGC 6624

NGC 6569

NGC 6907

62

M70   M69

N 6

ψ

NGC 6652

ε

ω

M55

η   NGC 6563

IC 5039

NGC 6726   NGC 6723

−30°
21h

NGC 6925

θ²

γ   ε

NGC 6729

λ   κ

θ¹

SAGITTARIUS

α

β

μ

β

IC 5013

δ

θ

α

IC 1297

ζ

NGC 6958

η²

κ¹

ι

η¹

NGC

κ²

α

CORONA
AUSTRALIS

MICROSCOPIUM

ζ

β¹

IC 4808

δ²   α

ε

β²

δ¹

η

ι

ι

ζ

θ¹

−40°

TELESCOPIUM

θ²

NGC 6868

κ

NGC 6

ζ   α

ρ

INDUS

λ

SOUTH EQUATORIAL 5

This map shows constellations and deep-sky
objects during mid-summer in the Northern
Hemisphere and mid-winter south of the equator.

142

Constellation boundary	Constellation figure	Magnitudes	
		Sirius	4.0
		0.0	5.0
		1.0	6.0
		2.0	7.0
		3.0	8.0

Star Chart — Scorpius / Ophiuchus / Lupus / Norma / Ara / Centaurus region

Top axis (Right Ascension): 18h · 17h · 16h

Right axis (Declination): −20° · −30° (15h) · −40°

Bottom axis: 18h · 17h · 16h · −50° · 15h

Map 19 →

Constellations
- OPHIUCHUS
- LIBRA
- SCORPIUS
- LUPUS
- NORMA
- ARA
- CENTAURUS

Labeled objects and stars

NGC 6356, M9, NGC 6342, TW, η, χ, φ, ψ, ν, β, ω¹, ω², θ, η, γ, ζ, κ, λ, τ

ξ, NGC 6369, NGC 6287, NGC 6235, NGC 6401, 44, o, θ, NGC 6284, ρ, ω, M80

NGC 6293, M19, ι, NGC 6144, Antares, σ, δ, o, π · 2, Me 2-1

NGC 6316, α, M4, NGC 6304, 45, Hb 5, M62, τ

M6, Tr 28, NGC 6383, CRL 6815, ρ, υ, τ

NGC 6416, NGC 6357, ε, χ, ξ, ψ¹, ψ²

M7, NGC 6334, NGC 6441, λ, υ, NGC 6302, NGC 6072, φ¹, φ², NGC 5824

Shaula, Ton 2, κ, NGC 6380, NGC 6337, μ, NGC 6139, θ, η, NGC 5986, υ

ι², ι¹, NGC 6153, NGC 6124, γ

NGC 6496, θ, η, ζ, NGC 6231, NGC 6192, λ, ω, δ

NGC 6388, NGC 6259, μ, γ, κ

σ, ι, NGC 6250, δ, θ, HJ 4788, ε, β, η

NGC 6352, NGC 6193, ε, θ, NGC 5882, o, λ

λ, IC 4651, α, NGC 6188, NGC 6134, ν¹, μ, ESO 274-1

κ, NGC 6167, η, γ², γ¹, ν², π, NGC 5643

μ, NGC 6152, Menzel 3

NGC 6397, ε², ε¹

M23, NGC 6445, NGC 6440

Legend
- ⬡ Open cluster
- ⊕ Globular cluster
- □ Diffuse nebula
- ✦ Planetary nebula
- ⬭ Galaxy

Greek alphabet
α Alpha	η Eta	ν Nu	τ Tau
β Beta	θ Theta	ξ Xi	υ Upsilon
γ Gamma	ι Iota	o Omicron	φ Phi
δ Delta	κ Kappa	π Pi	χ Chi
ε Epsilon	λ Lambda	ρ Rho	ψ Psi
ζ Zeta	μ Mu	σ Sigma	ω Omega

MAP **21** SOUTH EQUATORIAL 6

DANIEL VERSCHATSE

▲ **THREE MEMBERS** of the Grus Quartet (NGC 7582, right; NGC 7590, upper left; and NGC 7599) lie within 10' of one another. The fourth member (not shown), NGC 7552, lies ½° to the southwest.

Southern galaxies

The last of the mid-southern star maps features most of Capricornus the Sea Goat, Microscopium the Microscope, Piscis Austrinus the Southern Fish, and Grus the Crane — the only star figure in this region that resembles its namesake. We've also passed the Milky Way, so many more galaxies appear to populate this region.

Microscopium houses several galaxies that will test your observing prowess through an 8-inch telescope. Look for **IC 5039**. At magnitude 12.6, this object won't blind you, but it's noteworthy because of its interaction with IC 5041, a similarly bright galaxy located only 10' to the north-northeast. IC 5039 appears large with a broad core, while IC 5041's central region appears more concentrated than that of its neighbor.

Move 3° southwest of the galactic pair, and you'll find **NGC 6925**, a magnitude 11.3 spiral galaxy that appears three times as long as wide.

With a 12-inch scope, you can pick out the irregularly illuminated halo surrounding the moderately bright central region. The core appears stellar.

At magnitude 11.3, **NGC 6958** and NGC 6925 are the brightest galaxies in Microscopium. NGC 6958 measures 2.5' by 2' with a core slightly brighter than its outer regions. NGC 6925 measures 4.4' by 1.1' and marks the eastern corner of an equilateral triangle 3' on a side. Two 10th-magnitude stars sit at the other corners.

In Grus, several spiral galaxies are worth a look. Center Alpha Gruis in your eyepiece's field of view. Only 16' to the southeast lies **NGC 7213**. Luckily, this magnitude 10.0 galaxy is bright with an evenly illuminated central region. NGC 7213 measures 4.8' by 4.2'.

Next, find the Grus Quartet, the brightest member of which is **NGC 7582**. This galaxy glows at magnitude 10.6 and measures 6.9' by 2.6'. Two other members of the quartet are NGC 7590 (magnitude 11.5) and NGC 7599 (magnitude 11.4). These galaxies, which have similar shapes and

Designation	Right ascension	Declination	Magnitudes	Separation
Pi Capricorni	20h27m	−18°12'	5.2, 8.8	3.4"
Omicron Capricorni	20h30m	−18°35'	6.1, 6.6	18.9"
Alpha Microscopii	20h50m	−33°46'	5.0, 10.0	20.5"
Theta Indi	21h20m	−53°26'	4.7, 7.2	6.3"
Zeta Capricorni	21h27m	−22°25'	3.9, 12.5	21.3"
Iota Piscis Austrini	21h45m	−33°01'	4.4, 11.4	20.0"
Delta¹ Gruis	22h29m	−43°29'	4.0, 12.8	5.6"
Gamma Piscis Austrini	22h53m	−32°52'	4.6, 8.1	4.2"
Delta Piscis Austrini	22h56m	−32°32'	4.3, 9.3	4.9"
DUN 246	23h07m	−50°41'	6.1, 6.8	8.7"
Delta Sculptoris	23h49m	−28°07'	4.6, 11.6	3.8"

LEISURELYSCIENTIST (WIKIMEDIA COMMONS)

▲ **GLOBULAR CLUSTER** M30 in Capricornus offers a nice break from the many galaxies in this region of sky. M30 glows at magnitude 6.9.

FRED CALVERT/ADAM BLOCK/NOAO/AURA/NSF

▲ **SPIRAL GALAXY** NGC 6907 in Capricornus sports wide spiral arms. At magnitude 11.1, however, you'll need a big scope to see them.

ATLAS IMAGE COURTESY OF 2MASS/UMASS/IPAC-CALTECH/NASA/NSF

▶ **SPIRAL GALAXY** NGC 6925 in Microscopium has outer regions that appear irregular, a brighter central area, and a stellar core.

sizes to NGC 7582, lie 10' to the northeast. The last member, magnitude 10.7 NGC 7552, sits 0.5° west-southwest of NGC 7582.

Magnitude 10.9 **NGC 7314** sits about 1° northwest of Epsilon Piscis Austrini. This spiral galaxy measures 4.2' by 1.7' and stretches in a north-south orientation. Smaller, fainter NGC 7313 lies only 5' west of NGC 7314, just off the southern tip of the brighter galaxy's outer region.

If you'd like a change from galaxy observing, move into eastern Capricornus and locate globular cluster **M30**. This magnitude 6.9 object measures 11' across. Its core is concentrated, and its outer regions appear irregular through 6-inch and larger telescopes. Through a 10-inch scope at 200x, you'll count 200 stars.

If your telescope measures 16 inches or larger, try for the NGC 7103 galaxy group, 1° north of M30. **NGC 7103** is the brightest member, and it glows at a magnitude of only 12.6. The others are fainter. Don't expect to see detail in these objects — it's a victory just to spot them.

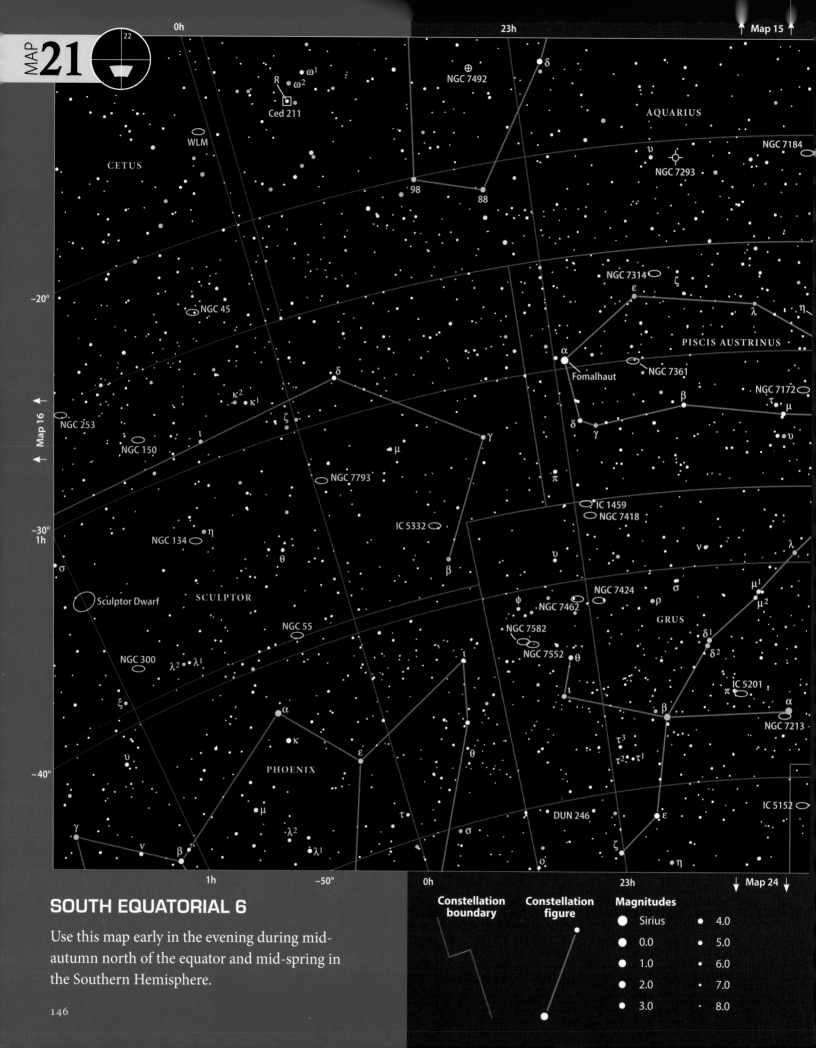

MAP 22 21

0h

23h

↑ Map 15 ↑

CETUS

AQUARIUS

ω¹
R
ω²
Ced 211

NGC 7492

NGC 7184

WLM

NGC 7293

υ

98

88

NGC 7314 ζ

−20°

ε

NGC 45

λ

η

PISCIS AUSTRINUS

α
Fomalhaut

NGC 7361

NGC 7172
τ μ

κ²
κ¹

δ

β

ζ
ι

γ
δ
γ

υ

NGC 253

μ

π

ν
λ

NGC 150

NGC 7793

IC 1459
NGC 7418

IC 5332

−30°
1h

η
NGC 134

θ

β

υ

σ

μ¹
μ²

SCULPTOR

Sculptor Dwarf

NGC 7424
σ
ρ

GRUS

φ
NGC 7462

δ¹

NGC 55

NGC 7582
NGC 7552 θ

δ²

NGC 300

λ² λ¹

IC 5201
π

ξ

ι

α
NGC 7213

α

β

υ

κ

θ

ι

τ³
τ² τ¹

−40°

PHOENIX

ε

IC 5152

μ

τ

DUN 246

ε

λ²

σ

γ

λ¹

ζ

ν
β

o

η

1h

−50°

0h

23h

↓ Map 24 ↓

SOUTH EQUATORIAL 6

Use this map early in the evening during mid-autumn north of the equator and mid-spring in the Southern Hemisphere.

Constellation boundary	Constellation figure	Magnitudes	
		⬤ Sirius	· 4.0
		⬤ 0.0	· 5.0
		● 1.0	· 6.0
		● 2.0	· 7.0
		● 3.0	· 8.0

CAPRICORNUS

Pal 12

NGC 7103

M30

24

ω

ψ

RT

M75

NGC 6907

NGC 6818

NGC 6822

θ

ι

κ

ε

η

φ

χ

ζ

δ

ε

γ

β

α

NGC 7135

γ

ι

IC 5148

θ² θ¹

ξ

η

IC 5039

NGC 6925

IC 5013

NGC 6958

ζ

θ² θ¹

MICROSCOPIUM

τ

ι

ζ

ν

κ¹

κ²

62

59

ω

52

χ³

χ¹

ψ

Map 20

M55

θ²

θ¹

SAGITTARIUS

ι

α

IC 1297

γ

α

β

δ

ζ

NGC 6726

NGC 6723

ε

NGC 6729

CORONA
AUSTRALIS

μ

NGC 7041

NGC 7049

INDUS

α

NGC 6868

TELESCOPIUM

β¹

β²

η²

η¹

θ

IC 4808

ι

θ

ι

η

Abell 3716

-20°

-30°

19h

-40°

22h

21h

20h

-50°

19h

	Open cluster	α	Alpha	η	Eta	ν	Nu	τ	Tau
	Globular cluster	β	Beta	θ	Theta	ξ	Xi	υ	Upsilon
	Diffuse nebula	γ	Gamma	ι	Iota	ο	Omicron	φ	Phi
	Planetary nebula	δ	Delta	κ	Kappa	π	Pi	χ	Chi
	Galaxy	ε	Epsilon	λ	Lambda	ρ	Rho	ψ	Psi
		ζ	Zeta	μ	Mu	σ	Sigma	ω	Omega

MAP **22** SOUTH POLAR 1

DYLAN O'DONNELL

▲ **THE TARANTULA NEBULA (NGC 2070), which is part of the Large Magellanic Cloud, is the only extragalactic nebula visible to the naked eye. The prominent star cluster near the bottom of this image is NGC 2100.**

Clouds of Magellan

The next star map contains two spectacular objects — the **Large Magellanic Cloud** (LMC) in Dorado and the **Small Magellanic Cloud** (SMC) in Tucana. They received their names because one of the first Northern Hemisphere inhabitants to describe the clouds was Portuguese explorer Ferdinand Magellan in 1519, during his circumnavigation of the world.

These objects are not earthbound clouds, however, but satellite galaxies of the Milky Way. The LMC lies 170,000 light-years away, and the SMC is 210,000 light-years distant.

Magnitudes for such extended objects don't mean much, but the LMC shines at approximately 0 magnitude. So, if you took the light from Vega (Alpha Lyrae) and spread it out over an area equal to that of the LMC, it would look about the same. The SMC glows at magnitude 2.3.

The full extent of these galaxies as we see them in the sky is gigantic. The LMC occupies an area roughly 11° by 9°. More than 470 Full Moons would be needed to cover that much sky. The SMC covers a smaller, but still impressive, area — 4.5° by 2.5° (53 Full Moons).

Although the LMC and SMC are small irregular galaxies, many celestial objects within and around them are worth pulling out a telescope for. In the LMC, for example, a 1,000-light-year-wide star-forming region of glowing hydrogen known as the **Tarantula Nebula** (NGC 2070) provides an easy target for telescopes of all sizes.

Near the SMC (but belonging to the Milky Way) is the brilliant globular cluster **47 Tucanae** (NGC 104). This magnitude 4.0 object (the second-brightest globular) lies 13,400 light-years from Earth and covers approximately the same amount of sky as the Full Moon — ½°.

But the Tarantula Nebula and 47 Tucanae are just the beginning. The NGC and IC catalogs tally nearly 400 deep-sky objects in the LMC and an additional 37 in the SMC. To see many of them, you'll need a big telescope, and you'll have to use high magnification. But even through a small telescope or binoculars, these two galaxies will fascinate you.

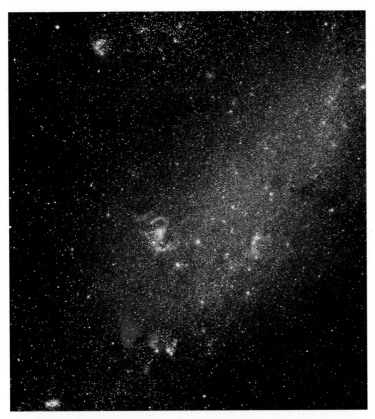

▲ THE SMALL MAGELLANIC CLOUD — a satellite galaxy of our Milky Way — covers an area equal to 53 Full Moons. To its right, globular cluster NGC 104, also known as 47 Tucanae, is part of our galaxy.

◀ THE LARGE MAGELLANIC CLOUD is the largest of the Milky Way's satellite galaxies. It lies 170,000 light-years away within Dorado and Mensa and shines as brightly as a magnitude 0 star. DYLAN O'DONNELL

▲ BARRED SPIRAL GALAXY NGC 2422 shines at magnitude 11.3 and lies 50 million light-years away in Volans the Flying Fish. It appears only 5' across, so use a large scope and high magnification.

/// DOUBLE-STAR DELIGHTS — MAP 22

Designation	Right ascension	Declination	Magnitudes	Separation
HJ 3435	1h25m	−59°29'	7.0, 9.0	25.5"
COO 14	2h39m	−52°57'	7.4, 8.3	8.8"
Gamma Horologii	2h45m	−63°42'	5.7, 13.0	20.0"
HJ 3568	3h08m	−78°59'	5.7, 9.4	15.2"
Epsilon Reticuli	4h17m	−59°17'	4.4, 12.5	13.7"
Theta Reticuli	4h18m	−63°15'	6.2, 8.2	4.0"
RMK 4	4h24m	−57°04'	7.1, 7.5	5.7"
Iota Pictoris	4h51m	−53°27'	5.6, 6.4	12.5"
Eta¹ Pictoris	5h03m	−49°08'	5.4, 13.0	10.6"
HDO 192	5h30m	−63°55'	6.3, 11.4	9.2"
HJ 3911	6h48m	−76°51'	6.9, 10.4	21.8"
Gamma Volantis	7h09m	−70°29'	3.9, 5.8	13.6"
Zeta Volantis	7h42m	−72°36'	3.9, 9.7	16.7"

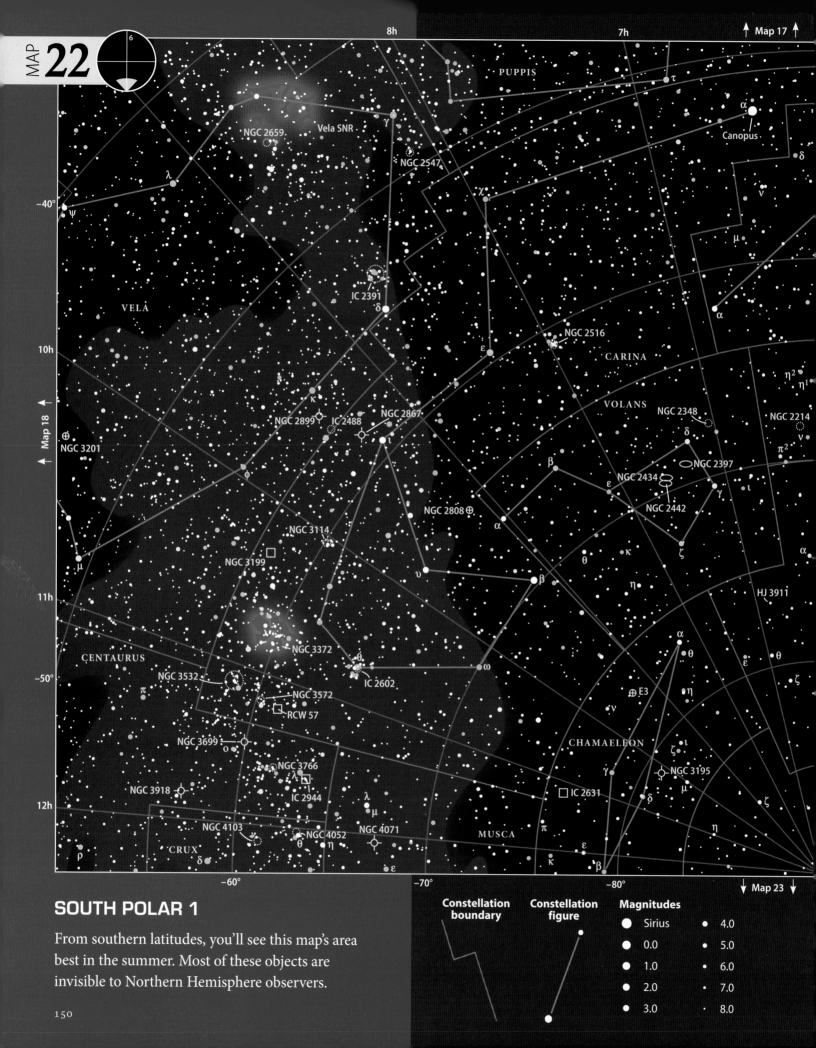

MAP **22** 6

8h

7h

↑ Map 17

PUPPIS

NGC 2659

Vela SNR

NGC 2547

γ

τ

α

Canopus

δ

NGC 2516

χ

ν

IC 2391

δ

ε

CARINA

μ

−40°

ψ

λ

VELA

10h

κ

NGC 2899

IC 2488

NGC 2867

ι

VOLANS

NGC 2348

η²

η¹

NGC 2214

δ

π²

ν

φ

⊕
NGC 3201

Map 18

β

ε

NGC 2397

γ

ι

α

NGC 2434

NGC 2808 ⊕

α

NGC 2442

NGC 3114

υ

θ

κ

ζ

α

NGC 3199

HJ 3911

11h

η

β

α

θ

ε

θ

CENTAURUS

NGC 3372

θ

ω

ζ

IC 2602

⊕ E3

η

NGC 3532

−50°

π

NGC 3572

ν

RCW 57

CHAMAELEON

ι

NGC 3699

o

ζ

NGC 3766

γ

NGC 3195

μ

NGC 3918

IC 2944

λ

δ

ζ

12h

μ

IC 2631

η

NGC 4103

π

NGC 4052

NGC 4071

MUSCA

CRUX

θ

η

ε

ε

ρ

δ

ε

κ

β

−60°

−70°

−80°

↓ Map 23 ↓

SOUTH POLAR 1

From southern latitudes, you'll see this map's area
best in the summer. Most of these objects are
invisible to Northern Hemisphere observers.

Constellation boundary	Constellation figure	Magnitudes	
		⬤ Sirius	• 4.0
		⬤ 0.0	• 5.0
		⬤ 1.0	· 6.0
		⬤ 2.0	· 7.0
		• 3.0	· 8.0

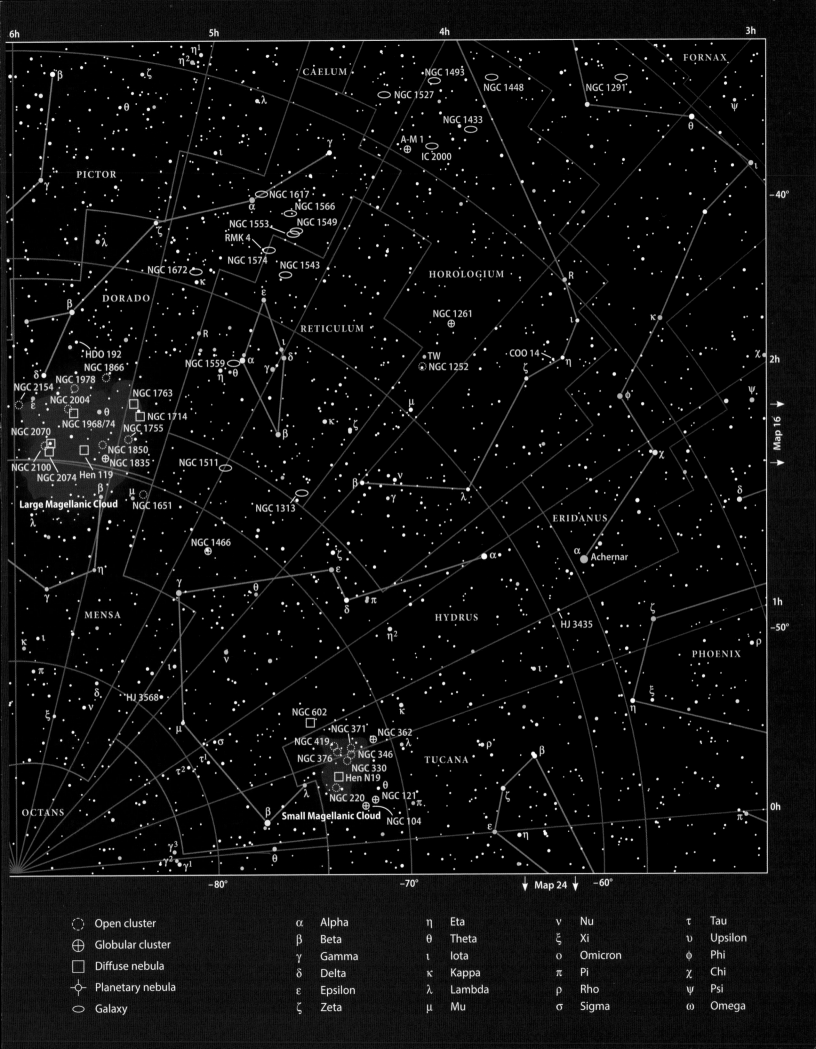

FORNAX

CAELUM

NGC 1493

NGC 1527

NGC 1448

NGC 1291

NGC 1433

ψ

NGC 1261

NGC 1617

NGC 1566

θ

ι

A-M 1

IC 2000

PICTOR

γ

NGC 1553

NGC 1549

−40°

RMK 4

NGC 1574

NGC 1543

ζ

λ

HOROLOGIUM

R

NGC 1672

κ

DORADO

β

R

ε

RETICULUM

ι

COO 14

κ

2h

HDO 192

NGC 1866

δ

NGC 1559

ι

δ

TW

NGC 1252

ζ

η

φ

NGC 2154

NGC 1978

ε

NGC 2004

θ

NGC 1763

α

η

θ

γ

μ

χ

ψ

NGC 1968/74

NGC 1714

κ

ζ

NGC 2070

NGC 1755

χ

NGC 1850

NGC 1835

NGC 2100

Hen 119

NGC 1511

ν

ERIDANUS

δ

NGC 2074

β

NGC 1313

β

γ

λ

Large Magellanic Cloud

μ

NGC 1651

α

Achernar

λ

NGC 1466

η

HJ 3435

ζ

1h

γ

ζ

ε

γ

θ

θ

δ

π

−50°

MENSA

δ

HYDRUS

η²

PHOENIX

κ

ι

ρ

π

ι

ξ

ν

δ

κ

η

HJ 3568

μ

NGC 602

σ

NGC 371

NGC 362

τ¹

NGC 419

λ

τ²

NGC 376

NGC 346

TUCANA

β

NGC 330

ρ

Hen N19

ζ

θ

NGC 220

NGC 121

λ

π

OCTANS

β

NGC 104

ε

Small Magellanic Cloud

η

γ³

γ²

γ¹

θ

π

Map 16 →

Map 24 ↓

◌ Open cluster	α Alpha	η Eta
⊕ Globular cluster	β Beta	θ Theta
☐ Diffuse nebula	γ Gamma	ι Iota
✧ Planetary nebula	δ Delta	κ Kappa
⬭ Galaxy	ε Epsilon	λ Lambda
	ζ Zeta	μ Mu

ν Nu	τ Tau
ξ Xi	υ Upsilon
ο Omicron	φ Phi
π Pi	χ Chi
ρ Rho	ψ Psi
σ Sigma	ω Omega

MAP 23   SOUTH POLAR 2

▲ **GLOBULAR CLUSTER** NGC 2808 in Carina shines at magnitude 6.2, making it visible to the naked eye from a dark site. The double star to the right of the cluster has reddish components of magnitudes 10.0 and 10.6.

Southern jewels

The accompanying map contains plenty to observe, including all of Crux the Southern Cross, the smallest of the sky's 88 constellations. You can see something interesting just looking at Crux without optical aid. Near its brightest star, Alpha Crucis, is a dark, 4°-wide swath of sky called the **Coalsack**. Only one 5th-magnitude star intrudes here.

On the eastern edge of Crux, the **Jewel Box Cluster** (NGC 4755) also is visible to the naked eye. This clump of colorful stars contains at least 25 members brighter than 12th magnitude and looks great through any size telescope. Three bright stars — one yellow, one blue, and one orange — form a line across the cluster's center. The rest of the stars are white and form a sparkly background to the three luminaries.

Planetary nebula **NGC 4071** lies 2° east-southeast of Lambda Muscae. Long-exposure images show it is an ellipse flanked by bright ends and containing a bar that crosses the nebula's minor axis. Through a 12-inch or larger telescope, NGC 4071 appears like a ghostly oval bubble about 1' across with a diffuse edge. A 13th-magnitude star appears near the center, but this is not the planetary's true central star, which is 19th magnitude. An OIII filter is a must if you want to see any of this object's details.

Only 0.7° southwest of Gamma Muscae lies **NGC 4372**, one of the least concentrated globular clusters. At a distance of 15,000 light-years, its brightest stars glow at 12th magnitude and are visible easily through small telescopes. An 8-inch scope shows NGC 4372 as a loose collection of

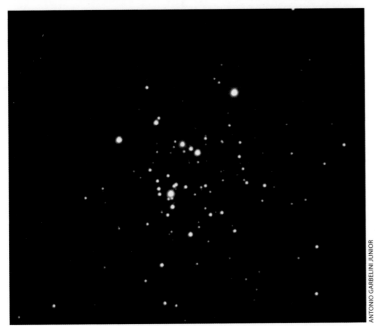

▲ **THE JEWEL BOX CLUSTER** (NGC 4755) is a beautiful celestial object through any size telescope. This 4th-magnitude open cluster is 10' across.

▲ **THE COALSACK** appears to stamp out the stars in the bright plane of our galaxy. It spans roughly 4° adjacent to Crux the Southern Cross.

/// DOUBLE-STAR DELIGHTS — MAP 23

Designation	Right ascension	Declination	Magnitudes	Separation
RMK 8	8h15m	−62°54'	5.3, 8.0	3.9"
HJ 4156	8h55m	−60°39'	4.0, 12.8	21.1"
HJ 4206	9h17m	−74°53'	5.5, 10.0	7.1"
Upsilon Carinae	9h47m	−65°03'	3.1, 6.1	5.0"
DUN 94	10h39m	−59°11'	4.8, 8.2	14.5"
Omicron¹ Centauri	11h32m	−59°26'	5.0, 11.4	13.5"
Lambda Centauri	11h36m	−63°01'	3.3, 11.5	16.3"
Gamma Crucis	12h32m	−57°06'	1.6, 6.7	10.6"
Alpha Muscae	12h37m	−69°08'	2.9, 13.0	29.6"
Iota Crucis	12h46m	−60°58'	4.7, 9.5	27.0"
Theta Muscae	13h08m	−65°18'	5.9, 7.5	5.3"
Alpha Circini	14h43m	−64°57'	3.4, 8.8	15.7"

13th- and 14th-magnitude stars spread over an area 19' across.

This cluster shows almost no concentration toward its center; at 150x, it looks like a circular open cluster. Through a 14-inch or larger scope, you can resolve NGC 4372 to the core, and it becomes a magnificent collection of mostly equal-magnitude stars with no background glow.

A more typical globular cluster is nearby **NGC 4833**, a nice contrast to NGC 4372. This is a moderately concentrated ball of stars that resolves well through 6-inch scopes. Use a 10-inch telescope, and you'll see dozens of 13th-magnitude cluster stars. Unlike NGC 4372, the broadly concentrated core displays a haze of unresolved stars.

The central region of NGC 4833 is its most concentrated area.

It appears like an oval stretched both east and west. This globular shines at magnitude 7.8, measures 13.5' across, and lies 25,000 light-years away.

Don't miss 10th-magnitude **NGC 5189** some 5.8° northeast of NGC 4833. This planetary nebula is a real oddball. At 5,000 light-years away, NGC 5189 is 2' across. Sometimes called the Barred Spiral Nebula because of its resemblance to a barred spiral galaxy through small telescopes, this planetary displays an S shape that's hard to miss.

Be sure to explore the small constellation Chamaeleon because it contains one of the highlights of the south polar region: planetary nebula **NGC 3195**. Although it glows at a paltry magnitude 11.6, this object's high surface brightness allows you to use high magnification to study it.

If your sky permits, use at least 200x on the 40" disk to reveal a slight north-south elongation and subtle brightness differences along NGC 3195's minor axis. The central star — listed at magnitude 15.3 — is a test only for the largest amateur telescopes because of the nebula's brightness.

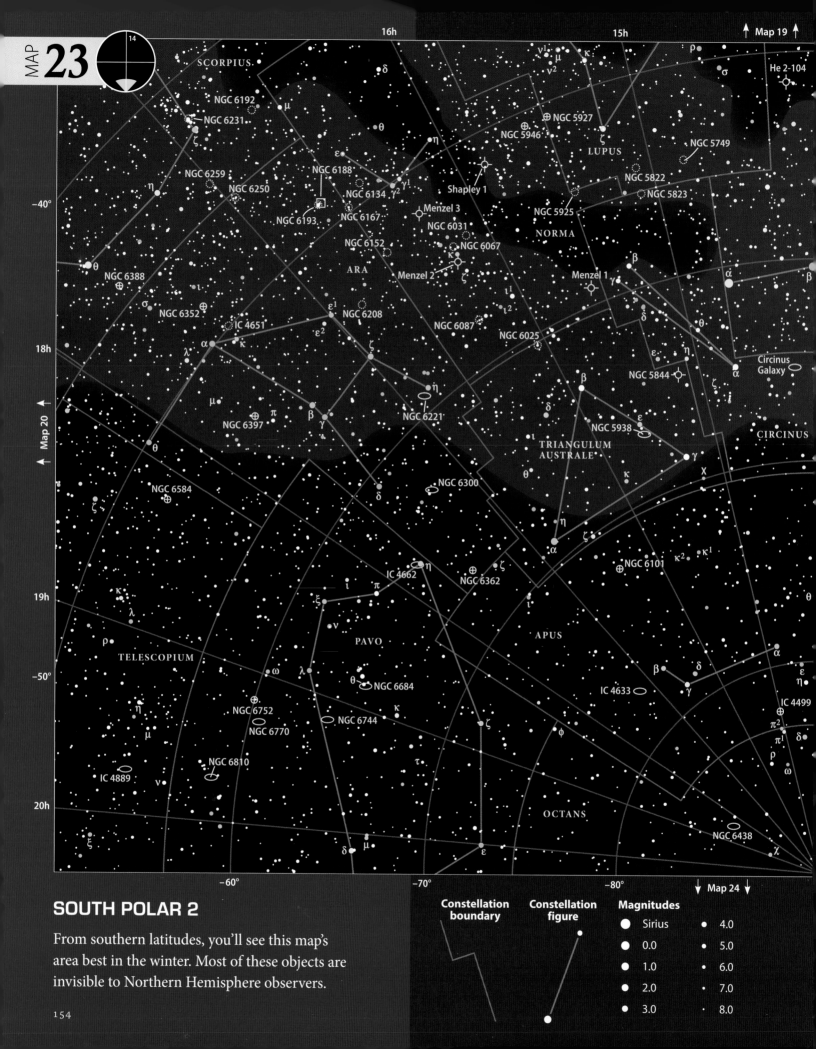

MAP
23 14

16h 15h ↑ Map 19 ↑

SCORPIUS

NGC 6192

NGC 6231

ζ

NGC 6259

η

NGC 6250

θ

NGC 6388

σ

NGC 6352

λ α κ
IC 4651
μ
NGC 6397 π
β
γ

θ

NGC 6584

ζ

δ

μ

δ

NGC 6188

NGC 6134 γ²
γ¹

NGC 6193

NGC 6167

NGC 6152

ARA

ε¹
NGC 6208
ε²
ζ

η
NGC 6221

Shapley 1

Menzel 3

NGC 6031

NGC 6067

Menzel 2 κ
ζ

NGC 6087

ν¹ μ κ
ν²

NGC 5946

ζ

LUPUS

NORMA

ι¹
ι²

NGC 6025

ρ
σ
He 2-104

NGC 5927

NGC 5749

NGC 5822

NGC 5823

NGC 5925

β

Menzel 1 γ

δ

ε

α β

α

θ

η

NGC 5844

β
δ

ι

TRIANGULUM
AUSTRALE
θ

NGC 6300

η
ζ

α

Circinus
Galaxy

ζ
χ

ε

NGC 5938

CIRCINUS

γ

κ

κ² κ¹

NGC 6101

IC 4662 η

π
ξ

ι

ν

PAVO

θ
NGC 6684

NGC 6752

NGC 6744

κ

NGC 6770

NGC 6810

IC 4889
ν

ζ

NGC 6362

ζ
ι

APUS

β δ
γ
IC 4633

φ

τ

μ
δ

ε

OCTANS

α
η
IC 4499
π²
π¹ δ
ρ
ω

NGC 6438
χ

θ

Map 20

TELESCOPIUM

κ
λ

ρ

η
μ

ξ

18h

19h

-50°

20h

ω λ

ξ

↓ Map 24 ↓

-40°

θ ι

-60° -70° -80°

SOUTH POLAR 2

From southern latitudes, you'll see this map's
area best in the winter. Most of these objects are
invisible to Northern Hemisphere observers.

Constellation boundary	Constellation figure	Magnitudes	
		● Sirius	
		● 0.0	● 4.0
		● 1.0	• 5.0
		● 2.0	• 6.0
		● 3.0	· 7.0
			· 8.0

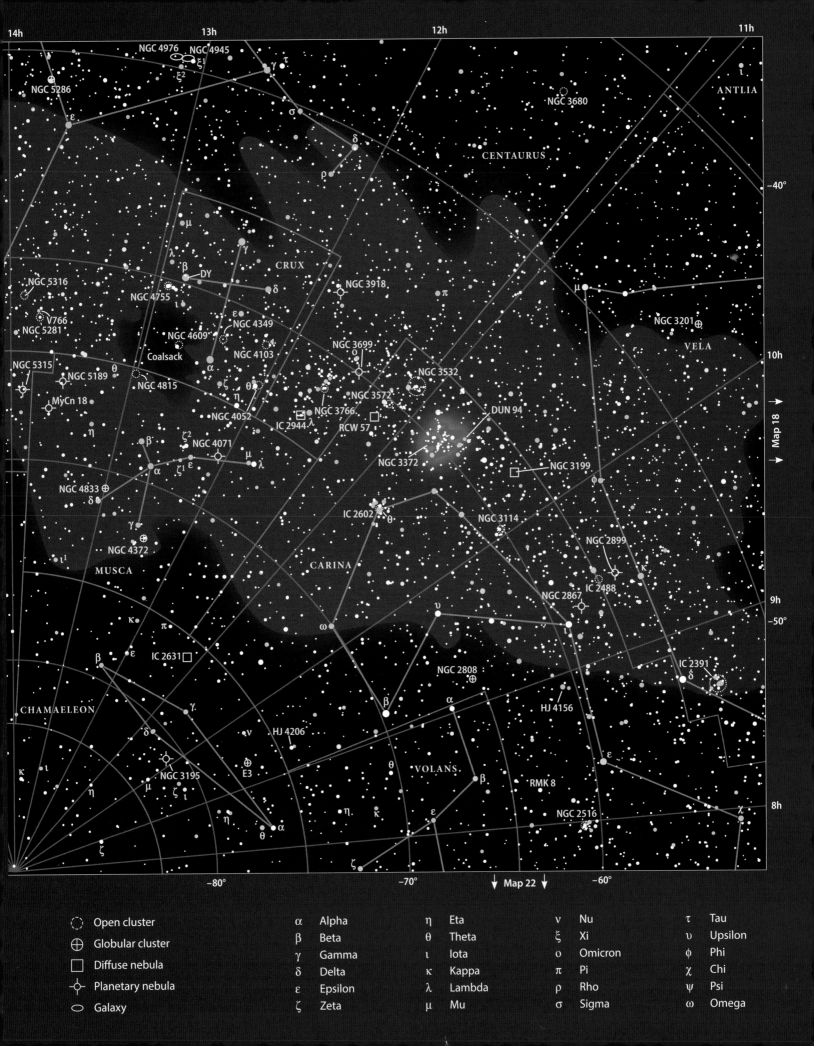

MAP **24** SOUTH POLAR 3

Faint southern stars

▲ **GLOBULAR CLUSTER** NGC 6397 in Ara is visible to the naked eye from a dark site. This cluster is a good one to observe through small telescopes because of the many 10th-magnitude stars it contains. MARC VAN NORDEN

The region of Map 24 at first seems to present a challenge. The constellations are faint, and only one 1st-magnitude star lights the way. Still, with a good telescope under a clear, dark sky, you can observe many treasures.

Point your telescope to the western edge of Pavo, near magnitude 3.6 Eta Pavonis, and you'll find the irregular galaxy **IC 4662** just 10' northeast of the star. For a better view of the magnitude 11.3 galaxy, crank up the magnification and position Eta Pav just outside the field of view. Through a 12-inch or larger telescope, you'll be able to pick out IC 4662's large stellar association that contains two emission nebulae. This complex dominates the galaxy's appearance. IC 4662 measures 3.2' by 1.9'.

NGC 6684, a magnitude 10.4 barred spiral galaxy, lies 6° east of Eta Pav and just 10' south of Theta Pavonis. Through an 8-inch scope, you'll see a bright core surrounded by a circular halo 2' across.

Move another 3° to the east-northeast, and you'll find spiral galaxy **NGC 6744**. With a magnitude of 8.6, you might think this would be a spectacular galaxy with plenty of spiral detail. Unfortunately, that's not the case. Because the light from NGC 6744 spreads out over a large area, small and medium telescopes don't bring the spiral arms into view. The magnitude 8.0 star SAO 254453 lies 0.6° west of NGC 6744.

Through a 12-inch telescope, you'll see a bright oval measuring 5' by 3' with an even brighter core. The outer reaches of the halo appear clumpy,

which suggests a spiral structure, but the arms are not easy to see.

Just 4° north of NGC 6744 lies one of the sky's brightest globular clusters, **NGC 6752**. You'll have no trouble spotting this magnitude 5.3 object with your naked eyes from a dark site. Through a 6-inch telescope, you'll see hundreds of stars starting at the cluster's outer regions and continuing to the strongly concentrated core. An 8th-magnitude star lies near the southeastern edge of the densest part of the cluster. NGC 6752 looks big and bright because it's nearby, only 13,000 light-years away.

In the southern reaches of Pavo, you'll find the bright elliptical galaxy **NGC 6876** 2.5° northeast of 4th-magnitude Epsilon Pavonis. This object lies at the center of a rich galaxy cluster. A 10-inch telescope shows the evidence; it brings several fainter companions into view. NGC 6876 appears as a bright circular smudge with a fairly wide, concentrated core. Two 12th-magnitude companions, NGC 6877 and NGC 6880, lie 2' and 6' to the east, respectively. And panning 9' to the northeast will bring you to NGC 6872, a barred spiral galaxy that reveals only a faint core to all but the largest amateur telescopes.

Midway between Theta and Delta Indi lies the fine spindle of **NGC 7090**. At a distance of just 20 million light-years, this spiral galaxy extends a full 6' in an 8-inch scope. It's also reasonably bright, glowing at magnitude 10.7. The galaxy's disk inclines just 5° to our line of sight and

◀ **A GALACTIC TRIO** interacts gravitationally in the northern section of Pavo. Thus, NGC 6769 (upper right), NGC 6770 (upper left), and NGC 6771 contain many new blue stars and pink star-forming regions.

DANIEL VERSCHATSE

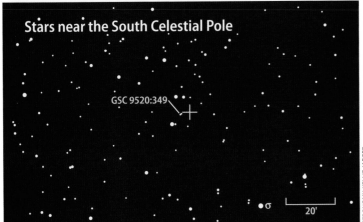

Stars near the South Celestial Pole

GSC 9520:349

σ   20'

ASTRONOMY: KELLIE JAEGER

▲ **THIS CHART** shows all stars near the South Celestial Pole (SCP, marked by the +) brighter than magnitude 11. Use this chart when you're south of the equator to polar-align your telescope. The star just to the lower left of the SCP is designated GSC 9520:349. The star nearest to the SCP usually visible to the naked eye is Sigma Octantis, which lies 1°3.5' below and to the right of the pole. Sigma Oct glows at magnitude 5.45.

/// **DOUBLE-STAR DELIGHTS — MAP 24**

Designation	Right ascension	Declination	Magnitudes	Separation
Kappa Tucanae	1h16m	−68°52'	5.1, 7.3	5.2"
HDO 262	17h00m	−54°35'	5.9, 12.0	20.1"
Gamma Arae	17h25m	−56°22'	3.5, 10.5	17.9"
HJ 4978	17h50m	−53°37'	6.0, 9.0	12.3"
Xi Pavonis	18h23m	−61°29'	4.3, 8.6	3.3"
HJ 5171	20h15m	−64°26'	7.0, 10.0	17.3"
HJ 5182	20h33m	−80°58'	5.8, 11.5	26.8"
Mu Octantis	20h42m	−75°21'	7.1, 7.6	17.4"
Beta Indi	20h55m	−58°27'	3.7, 12.5	17.3"
Theta Indi	21h20m	−53°26'	4.7, 7.2	6.0"
Lambda Octantis	21h51m	−82°43'	5.5, 7.8	2.8"
Delta Tucanae	22h27m	−64°58'	4.8, 9.3	6.9"

measures less than 1' wide. Use a 12-inch telescope, and you'll see a broad, concentrated center. Any further detail, such as mottling in the spiral arms, takes a much larger scope.

In Octans, you can find **NGC 7098** lying pretty much alone. Through a 12-inch telescope, the galaxy measures 3' by 2' and appears somewhat stretched on a northeast-to-southwest line. The galaxy is moderately bright (magnitude 11.4), and it has a broad, concentrated center. Look for a pair of 11th- and 12th-magnitude stars 6' to the southwest.

Remember Kemble's Cascade (see page 64)? A similar object — **Melotte 227** — lies in the southernmost constellation, Octans. Melotte 227 looks like an open cluster containing about 20 stars between magnitudes 7 and 10. In 1998, however, astronomers learned the stars don't share a common motion through space but are a random concentration of stars at various distances. Still, it's worth a look through binoculars or a telescope/eyepiece combination that gives at least a 2° field of view.

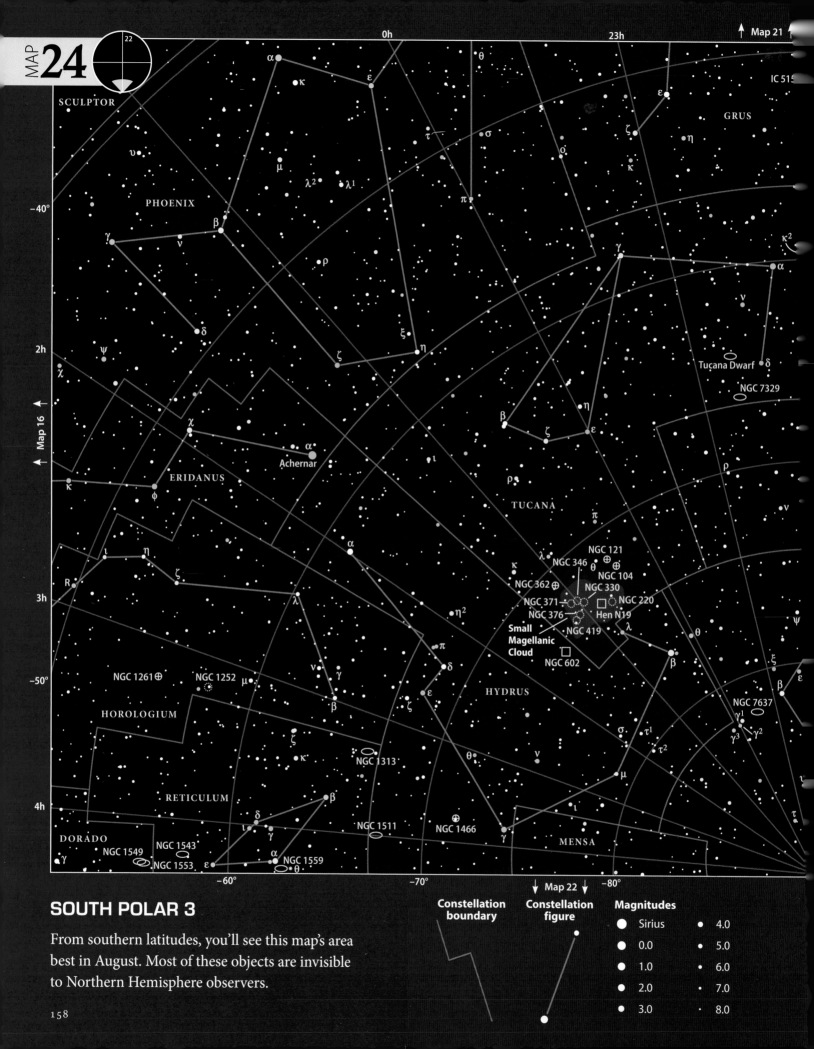

MAP
24
22

↑ Map 21 ↑

0h
23h

IC 515

SCULPTOR

GRUS

α
κ
ε
θ
ε

Tucana Dwarf

NGC 7329

-40°

PHOENIX

υ

μ
λ²
λ¹

τ
σ
π
o
ζ
η
κ

γ
κ²
α

β
ν
ν

δ

ρ

Map 16

2h

ψ
δ

ξ
η
ζ

γ

χ

η
β
ζ
ε

ρ

χ
α
Achernar

ι

ρ
ν

ERIDANUS

TUCANA

φ

π

κ

λ
NGC 121
NGC 346
θ
NGC 104

ι
η

R
ζ

α

NGC 362
NGC 330

NGC 371
NGC 220
ψ

3h
3h

λ

η²

NGC 376
Hen N19

NGC 419
λ

θ

NGC 1261
NGC 1252
μ

ν
γ

π

Small
Magellanic
Cloud
NGC 602

β
β

δ

β

-50°

HOROLOGIUM

ε

HYDRUS

NGC 7637
γ¹

ζ
ζ

NGC 1313

θ

σ
τ¹
γ³
γ²

κ

ν
τ²

μ

ι

RETICULUM

β

ι

4h

δ

NGC 1511
NGC 1466

γ

MENSA

ι

DORADO
γ

ι
γ

α

NGC 1549
NGC 1543
NGC 1559

τ

NGC 1553
ε
θ

-60°
-70°
↓ Map 22 ↓
-80°

SOUTH POLAR 3

From southern latitudes, you'll see this map's area
best in August. Most of these objects are invisible
to Northern Hemisphere observers.

**Constellation
boundary**

**Constellation
figure**

Magnitudes

Sirius
0.0
1.0
2.0
3.0

4.0
5.0
6.0
7.0
8.0

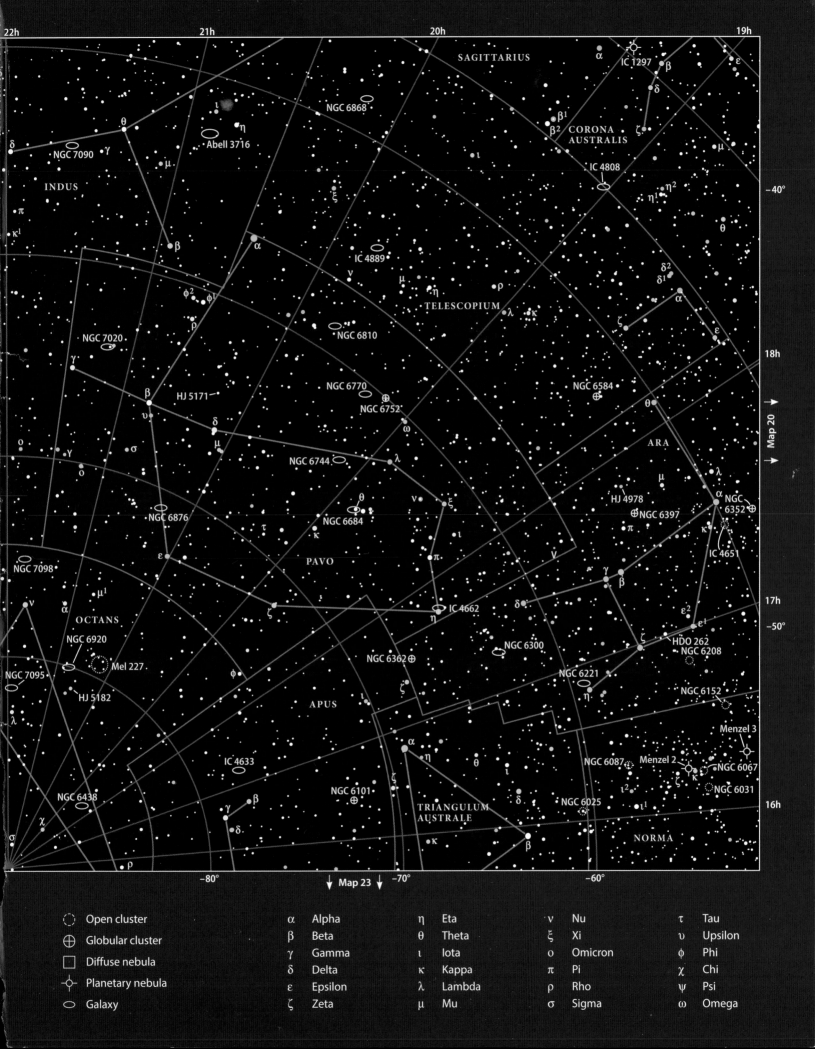

SAGITTARIUS

IC 1297

CORONA
AUSTRALIS

NGC 6868

−40°

θ
η
ι

Abell 3716

β¹
β²

INDUS

ζ

μ

NGC 7090

δ

γ

IC 4808

μ

η¹
η²

θ

μ
γ

ξ

δ²
δ¹

π

κ¹

β

α

IC 4889

ν

μ

ρ

α

ζ

ε

18h

φ²
φ¹

ρ

TELESCOPIUM

λ
κ

η

NGC 7020

NGC 6810

NGC 6584

θ

γ

HJ 5171

NGC 6770

ARA

Map 20

β
υ

δ
μ

NGC 6752

ω

μ

λ

HJ 4978

α
NGC
6352

NGC 6744

λ

NGC 6876

θ

ν

ξ

ι

NGC 6397

π

κ

τ

NGC 6684

κ

IC 4651

PAVO

π

γ

NGC 7098

ε

17h

μ¹

ζ

γ
β

ν

α

IC 4662

δ

OCTANS

η

ζ

ε²

NGC 6920

NGC 6300

ζ

HDO 262

−50°

ε¹

Mel 227

NGC 6208

NGC 7095

NGC 6362

NGC 6221

HJ 5182

φ

ζ

η

NGC 6152

APUS

ι

Menzel 3

IC 4633

α

η

NGC 6087

Menzel 2

NGC 6067

NGC 6438

θ

ι

ζ

κ

NGC 6031

χ

ζ

NGC 6101

ι²

γ

β

TRIANGULUM
AUSTRALE

δ

NGC 6025

ι¹

16h

σ

δ

κ

NORMA

ρ

β

⬭ Open cluster	α Alpha	η Eta	ν Nu	τ Tau
⊕ Globular cluster	β Beta	θ Theta	ξ Xi	υ Upsilon
□ Diffuse nebula	γ Gamma	ι Iota	ο Omicron	φ Phi
✧ Planetary nebula	δ Delta	κ Kappa	π Pi	χ Chi
⬭ Galaxy	ε Epsilon	λ Lambda	ρ Rho	ψ Psi
	ζ Zeta	μ Mu	σ Sigma	ω Omega

The Perfect Products to Complement A Starry Sky

My Science Shop is your one-stop source for some of the best astronomy resources available. Shop the best-selling science books, DVDs, globes, maps, toys and many more products curated by the editors of *Astronomy* and *Discover* magazine.

Astronomy 40° North Planisphere

Learn the constellations, or easily and quickly determine when a particular part of the sky will be visible with this 10.5" diameter Planisphere.

#81108 • $11.95

Portrait of the Milky Way Print

Admire a 39" x 25" revised version of Lomberg's classic Portrait of the Milky Way mural commissioned by the National Air and Space Museum of the Smithsonian Institution.

#81107 • $21.99

Mars Globe

Display this 12" desktop globe that shows an astounding 206 features of the planet Mars.

#81091 • $99.95

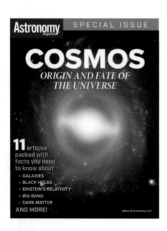

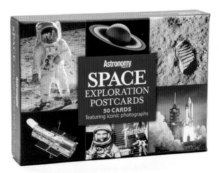

Cosmos: Origin and Fate of the Universe

Explore everything we know about how the universe originated and has evolved since The Big Bang 13.8 billion years ago.

#AS05200701-C • $9.99

Space Exploration Postcards

This collectible set of 50 Space Exploration Postcards features iconic photos of astronauts, Apollo missions, spacecraft and much more!

#81318 • $14.99

Constellation Flashcards

Learn about each constellation in the night sky: how to pronounce it, where and when to see it, fun facts, and more with this 36-card flashcard set.

#81077 • $10.99